NEUROSCIENCE RESEARCH PROGRESS

SPATIAL MEMORY: VISUOSPATIAL PROCESSES, COGNITIVE PERFORMANCE AND DEVELOPMENTAL EFFECTS

NEUROSCIENCE RESEARCH PROGRESS

Current Advances in Sleep Biology
Marcos G. Frank (Editor)
2009. ISBN: 978-1-60741-508-4

Neuroscience Research Advances
Benito Figueredo and Fidel Meléndez (Editors)
2009. ISBN: 978-1-60692-967-4

Dendritic Spines: Biochemistry, Modeling and Properties
Louis R. Baylog (Editor)
2009. ISBN: 978-1-60741-460-5

Relationship between Automatic and Controlled Processes of Attention and Leading to Complex Thinking
Rosa Angela Fabio
2009. ISBN: 978-1-60741-810-8

Neurochemistry: Molecular Aspects, Cellular Aspects and Clinical Applications
Anacleto Paços and Silvio Nogueira (Editors)
2009. ISBN: 978-1-60741-831-3

Sleep Deprivation: Causes, Effects and Treatment
Pedr Fulke and Sior Vaughan (Editors)
2009. ISBN: 978-1-60741-974-7

Color Perception: Physiology, Processes and Analysis
Darius Skusevich and Petras Matikas (Editors)
2009. ISBN: 978-1-60876-077-0

Translational Neuroscience in Animal Research: Advancement, Challenges, and Research Ethics
Jason E. Warnick and Allan V. Kalueff (Editors)
2010. ISBN: 978-1-60876-185-2

Developmental Neurotoxicity of PBDEs, Mechanisms and Implications
Henrik Alm and Birger Scholz
2010. ISBN: 978-1-61668-295-8
2010. ISBN: 978-1-61668-602-4 (E-book)

Spatial Memory: Visuospatial Processes, Cognitive Performance and Developmental Effects
Jacob B. Thomas (Editor)
2010. ISBN: 978-1-61668-139-5
2010. ISBN: 978-1-61668-467-9 (E-book)

Handbook of Neuropsychiatry Research
Rebecca S. Davies (Editor)
2010. ISBN: 978-1-61668-138-8
2010. ISBN: 978-1-61668-798-4 (E-book)

An Overview of Genetic Factors and Environmental Triggers in the Pathogenesis of Tourette's Syndrome
Cong-Yi Wang, Shi-Guo Liu, Ping Yang, Zhimin Miao, Qian Dong, Changgui Li and Wei Wang
2010. ISBN: 978-1-61668-748-9
2010. ISBN: 978-1-61728-254-6 (E-book)

Prenatal Alcohol Exposure and its Effects on Hippocampal Anatomy and Function
Joanna H. Sliwowska
2010. ISBN: 978-1-61668-922-3

Huntington's Disease: Etiology and Symptoms, Diagnosis and Treatment
Thomas J. Visser (Editor)
2010. ISBN: 978-1-61668-984-1
2010. ISBN: 978-1-61728-451-9 (E-book)

Neural Computation and Particle Accelerators: Research, Technology and Applications
Emmerich Chabot and Horace D'Arras (Editors)
2010. ISBN: 978-1-60741-280-9

Cultural and Linguistic Influence on Developmental Neural Basis of Theory of Mind: Whorfian Hypothesis Revisited
Chiyoko Kobayashi Frank
2010. ISBN: 978-1-60876-362-7

The Importance of Colour Perception to Animals and Man
Birgitta Dresp-Langley and Keith Langley
2010. ISBN: 978-1-60876-394-8

Learning and Memory Developments and Intellectual Disabilities
Lucas C. Eklund and Aron S. Nyman (Editors)
2010. ISBN: 978-1-60876-397-9

Dual Processing Model of Visual Information. Cortical and Subcortical Processing
Hitoshi Sasaki
2010. ISBN: 978-1-60876-399-3

Homeostatic Role of the Parasympathetic Nervous System in Human Behavior
Aurélien Pichon and Didier Chapelot
2010. ISBN: 978-1-60876-804-2

Rehabilitation of the Shaken Soldier Syndrome
Larry E. Schutz, David Patterson, Mathu Hanson and Elizabeth A. McNamara
2010. ISBN: 978-1-61668-192-0
2010. ISBN: 978-1-61668-492-1 (E-book)

NEUROSCIENCE RESEARCH PROGRESS

SPATIAL MEMORY: VISUOSPATIAL PROCESSES, COGNITIVE PERFORMANCE AND DEVELOPMENTAL EFFECTS

JACOB B. THOMAS
EDITOR

Nova Science Publishers, Inc.
New York

For permission to use material from this book please contact us:
Telephone 631-231-7269; Fax 631-231-8175
Web Site: http://www.novapublishers.com

LIBRARY OF CONGRESS CATALOGING-IN-PUBLICATION DATA

Available Upon Request

ISBN: 978-1-61668-139-5

Published by Nova Science Publishers, Inc. ✝ New York

CONTENTS

PREFACE

A fundamental requirement of everyday life is that of encoding and remembering successfully the locations of objects, landmarks or buildings in space. This function is achieved by structuring spatial information in systems of coordinates. It is often argued that a person's, or an animal's, spatial memories are summarized in a cognitive map. There are two main categories for classifying spatial reference systems. One is the egocentric frame of reference which specifies location and orientation with respect to the organism and include eye, head and body coordinates. Allocentric frames of reference specify location and orientation with respect to elements and features of the environment independently of the viewer's position. This new book presents and reviews important data on spatial memory.

Chapter 1- This chapter consists of five main parts. In the first one (I - Introduction) the role of cognitive memory in "becoming" of the complexly organized systems is briefly considered from the viewpoint of the general systems theory; the role of Georgian neurophysiologists in emphasizing cognitive aspects of memory role in animal behavior is underlined. In the second part (II - Nonspatial visual recognition memory) the existence of the "one-trial" nonspatial visual recognition memory in cats, as representatives of carnivores with poorly developed vision, is established; it is shown that analogously to primates and rodents in cats the so called rhinal cortex plays a major role in mediating this type of visual cognitive memory. The following part (III - Spatial visual recognition memory) is devoted to spatial visual "one-trial" memory in cats; it is shown that in cats, similarly to primates and rodents, parahippocampal cortex (with possible inclusion in this system of the subicular complex) is responsible for visuospatial memory. In the next to last part of this chapter (IV - Two subsystems of spatial visual memory – delayed response problem) two probable subsystems of visual spatial memory – egocentric and allocentric – are briefly reviewed in the classical delayed response context; it is shown that the first one depends on the prefrontal cortex, while the second one – on the parietal association region. The last part (V - Conclusion) emphasizes authors' persuasion that "mechanisms of recall" from the cognitive memory store might be studied most fruitfully just in the field of cognitive recognition memory because of its controlled nature in appropriate behavioral testing paradigms.

Chapter 2- A fundamental requirement of everyday life is that of encoding and remembering successfully the locations of objects, landmarks or buildings in space. This function is achieved by structuring spatial information in systems of coordinates. There are many ways to classify spatial reference systems (Paillard, 1991), but a useful one for understanding human spatial memory, divides them into two main categories. Egocentric frames of reference specify location and orientation with respect to the organism, and include eye, head, and body coordinates. Allocentric frames of reference specify location and orientation with respect to elements and features of the environment independently of the viewer's position.

Given the primary role that egocentric and allocentric processing systems play in perception and action (e.g. Milner & Goodale, 1995, 2008; but see Schenk, 2006), they can be thought of as specialized cognitive mechanisms essential in performing and guiding spatial motor actions. Egocentric representations have a special relevance in controlling movement in surrounding space such as avoiding obstacles or reaching objects. All these actions are performed in near/peripersonal space, i.e. the space within arm-reaching distance, and require fine-grained metric information. On the other hand, allocentric representations have an important role in recognizing objects, scenes and planning future movements (i.e. the space outside arm-reaching distance) (e.g. Milner & Goodale, 2008). Spatial memory is intrinsically linked to frames of reference as it is not possible to store spatial information without structuring it according to specific frames. In this chapter authors review evidence from behavioral and neurofunctional data about models of spatial memory, automatic and effortful encoding of spatial information, and spatial memory in elderly people. Furthermore, two experiments are reported that investigate memories for egocentric and allocentric frames of reference by means of a task requiring judgments of relative distance. Experiment 1 investigates the influence of near/far spaces on spatial frames of reference. Experiment 2 controls the influence of possible artifactual effects. Overall, the pattern of data shows that the egocentric processing is more accurate and faster than the allocentric one. The results are discussed in relation to models of spatial memory that emphasize the importance of egocentric experience (Kosslyn, 1994; Millar, 1994).

Chapter 3- There is abundant evidence that allocentric spatial memory declines with advancing age. However, spatial memory is not a unitary construct, but can be subdivided into memory for object locations and maze learning. Furthermore, performance on these functions may also be affected by age-associated decline in working memory, executive functioning (notably visuospatial planning) and long-term memory in general. It was the aim of the current study to investigate these aspects in combination in a group of younger (N = 25) and older (N = 25) adults. Maze learning was measured with a computer adaptation of the Morris Water Maze. Object-location memory function was measured using the Object Relocation paradigm, also including task conditions assessing object-location binding and positional memory. Furthermore, spatial working memory was measured with the Spatial Span subtest of the WMS-III, visual long-term memory with the Doors Test and visuospatial planning with the Key Search test. The results clearly showed large ($d > 0.8$) age effects on all aspects of object-location memory, spatial working memory and visual long-term memory, but not on visuospatial planning and maze learning. Also, spatial working-memory decline was highly correlated with the performance on all other tasks. These results suggest that spatial memory and spatial learning are differentially affected by age and may thus rely on distinct neurocognitive processes.

Chapter 4- One of the central aims in developmental cognitive neuroscience, is to relate changes in cognition to maturational changes in the brain. Research with animals has been helpful in understanding links between brain changes during development as they relate to cognition. A good example comes from studies of spatial memory in relation to the development of the hippocampus. Before 19 days of age, rats pups are unable to solve the hidden platform version of the Morris watter maze task. By 21 days of age, however, rats can typically solve the Morris water maze (Purves et al., 2008). In this structure were found a high density of steroid receptors, suggesting there must be some relationship between steroid receptors and development of hippocampus (Simerly et al., 1990).

Adolescence is now recognized as a major and dynamic period of neural development during which behavioral circuits are remodeled and refined. Although the human brain of a 5-year old child is already 90% of its adult size (Dekaban, 1978), significant remodeling is still to come. This concept was kindled by research in both humans and animals documenting that many of the basic developmental processes occurring during perinatal brain development are recapitulated during adolescence. Compared with activation of neural circuits by steroid hormones, it is relatively little known about steroid-dependent organization during adolescence. Presumably, there are many similarities to the processes involved in steroid-dependent sexual differentiation during early neural development, such as regulation of cell death and survival, and synaptic density and connectivity.

During the adolescent phase of organization, steroid-dependent refinement of neural circuits results in long-lasting structural changes in brain, which determine adult behavioral responses to hormones or sensory stimuli and improve the cognitive abilities such as attention, learning and memory, reasoning and visuospatial processing, as adaptative response to environment and social interaction.

The organizational changes in the adolescent brain under steroid hormones could evolve in a progressive manner, in order to respond from simple to complex stimuli and thus, the trajectory of cognitive development follow the developmental course of brain neuromaturation, from visual perception and spatial memory to visuospatial processing and semantic association between visual informations and abstract categories.

Chapter 5- How authors acquire and represent spatial information is one of the most important unsolved issues in spatial cognition. Siegel and White (1975) affirmed that different forms of environmental knowledge are acquired and represented depending on the type of information selected: *landmark*, characterized by environmental patterns that are perceptually salient or important for the person; *route*, based on the pathes generally used to connect landmarks; and *survey*, an overall configuration of the environment, similar to a map. According to these authors, anyone can reach survey representation with extensive experience of the environment. Recently, Pazzaglia and co-workers (2000) demonstrated that these three types of representations correspond to three different spatial cognitive styles. Conversely, according to Montello (1998) a pure landmark or route representation does not exist; indeed, during their first exposure to the environment, people acquire and represent an overall survey configuration of it.

Authors aim was to determine whether environmental familiarity and/or spatial cognitive style predict the way authors acquire and represent spatial information.

Forty participants who had different degrees of familiarity with the Italian city of Bologna took part in the experiment. Familiarity with Bologna was evaluated using a questionnaire. This city was selected because it has a small and well-defined centre that can be easily explored on foot. Participants were further subdivided by spatial cognitive style to assess its weight in environmental representation. They performed six spatial tasks concerning Bologna that measured different spatial abilities, based on Siegel and White's frameworks (1975).

Authors found that neither familiarity with the environment nor spatial cognitive style predict the correct solution of landmark tasks, whereas both familiarity with the environment and spatial cognitive style predict the correct solution of route and survey tasks.. Thus, authors can affirm that both familiarity with the environment and spatial cognitive style are important for acquiring and representing spatial information, but their involvement depends on task demands. To authors knowledge, this is the first study in which knowledge of a real town has been measured. This ecological setting allowed us to propose a new model to explain individual differences in moving successfully through the environment.

Chapter 6- Optics is considered to be the oldest branch of science as light and vision have long been considered to be the central for understanding the world in which we live. Since, ancient day's vision is regarded as the supreme sense for the acquisition of knowledge, and light is the key to unlock the secrets of the universe. The mechanism of vision is regarded as the formation of an image of the object we see, by the lens of the eye, on to a light sensitive-membrane in the back of the eye, called the retina. The retina serves as a transducer and converts the image pattern formed by the light into neuronal signals. These signals are processed in different parts of the brain and we finally see. However, the major problem in visual perception is that what people see is not simply a translation of the image on the retina. Thus people interested in perception have long struggled to explain what visual processing does to create what we actually see. Visual perception is the ability to interpret information from visible light reaching the eye. Hermann von Helmholtz [1] the pioneering man in the study of visual perception in modern times, felt that the poor quality of information gathered by the eye can not make vision possible and concluded that vision could only be the result of some form of unconscious inferences that is, to make assumptions and conclusions from incomplete data, based on inference data. Gestalt psychologists [2] have guided the study of how people perceive visual components as organized patterns instead of many different parts. Then onwards, the visual perception is considered a combination of physiological and psychological processes and different groups are thinking of many models for visual perception. During the last three decades many computational models of vision were developed which explained visual phenomena relatively closer to real life situation.

Chapter 7- Primate and human spatial location coding is based on a variety of different reference frames. Two fundamentally different types of reference system are usually considered, egocentric and allocentric. The first type codes locations with reference to the position of the body, or of specific parts of the body, the second codes locations with reference to the surrounding environment. Herein authors examine behavioural and neurophysiological evidence that suggests: (a) the existence of a series of co-ordinate transformations, from a retinotopic to body-centred spatial representation, for guiding bodily actions in space, and (b) the use of a body-centred co-ordinate system for coding, in short-term memory, of target locations to be reached with arm movements.

Chapter 8- Existing studies about the effects of attention on long-term memory encoding focused on the effects of 'frontal', executive attentional mechanisms and made use of paradigms of divided attention. Authors investigated the impact of visuo-spatial ('posterior') attention on long-term memory encoding. In four experiments, authors used a standard spatial cuing paradigm and presented words either under valid or invalid cuing conditions. Performance in yes-no-recognition, cued recall and free recall was enhanced in the valid cuing-condition, whereas no effect was found for source-monitoring. The results show not only divided, but visuo-spatial attention considerably influences memory encoding, and the impact of visuo-spatial attention is different from that of divided attention.

Chapter 9- This chapter investigates the nature of and association between visuospatial memory, and, separately, visuospatial working memory in adolescent onset schizophrenia. The data presented suggest that both the visuospatial memory and working memory performance of the adolescent-onset schizophrenia group are impaired compared to the healthy control group. Further, the type and magnitude of the deficits are similar to those reported in the adult-onset schizophrenia patients suggesting these deficits are developmental stage independent. The import of these findings is examined within the context of emerging neurobiological models of progressive parietal-temporal-frontal linked neural network dysfunction in schizophrenia, whether of adolescent- or adult-onset. Also, the implications of these findings for the current aetiological understanding of schizophrenia and the current assessment and treatment approaches are discussed.

In: Spatial Memory
Editor: Jacob B. Thomas, pp. 1-50
ISBN: 978-1-61668-139-5

Chapter 1

COGNITIVE VISUAL MEMORY IN CATS

V. M. Okujava[1] and T. A. Natishvili[2]

[1]Research Center for Experimental Neurology, Tbilisi, Georgia; and the Javakhishvili State University of Tbilisi, Tbilisi, Georgia

[2]Research Center for Experimental Neurology, Tbilisi, Georgia

ABSTRACT

This chapter consists of five main parts. In the first one (I - Introduction) the role of cognitive memory in "becoming" of the complexly organized systems is briefly considered from the viewpoint of the general systems theory; the role of Georgian neurophysiologists in emphasizing cognitive aspects of memory role in animal behavior is underlined. In the second part (II - Nonspatial visual recognition memory) the existence of the "one-trial" nonspatial visual recognition memory in cats, as representatives of carnivores with poorly developed vision, is established; it is shown that analogously to primates and rodents in cats the so called rhinal cortex plays a major role in mediating this type of visual cognitive memory. The following part (III - Spatial visual recognition memory) is devoted to spatial visual "one-trial" memory in cats; it is shown that in cats, similarly to primates and rodents, parahippocampal cortex (with possible inclusion in this system of the subicular complex) is responsible for visuospatial memory. In the next to last part of this chapter (IV - Two subsystems of spatial visual memory – delayed response problem) two probable subsystems of visual spatial memory – egocentric and allocentric – are briefly reviewed in the classical delayed response context; it is shown that the first one depends on the prefrontal cortex, while the second one – on the parietal association region. The last part (V - Conclusion) emphasizes authors' persuasion that "mechanisms of recall" from the cognitive memory store might be studied most fruitfully just in the field of cognitive recognition memory because of its controlled nature in appropriate behavioral testing paradigms.

I. Introduction

According to R. Gerard all complex material systems are characterized by three fundamental attributes: 1. Existence, 2. Behavior and 3. Becoming [Gerard, 1961]. "Existence" means simply being of system in the world during some period of time with stable structural-functional characteristics; "behavior" means rapid and reversible functional responses of the system to currently acting stimuli from the environment. Due to that attribute the system after responding to current environmental stimulation returns to its initial state. Last but not least, "becoming" means self-development of the system over time, due to acquisition of individual experiences, accumulation of information about environmental events and self-actions in that environment. At the heart of that fundamental attribute lie some subtle but frequently very stable material changes in the system which remains in it after stimulation. Of course we mean under such changes the engrams, constituting very bases of memory – quite important function of very complex, self-organized systems of different physical origin [Adey, 1967; Ashby, 1960; Von Neumann, 1951].

To-day not only computer scientists and psychologists are interested in the intimate nature and organization of memory system in living brain, but neurophysiologists as well, who possess acute sense that: "there exists a striking break in our knowledge of short-term changes in neural and muscular cells (impulse propagation, synaptic transmission, muscular contraction) and those long-term events in the neural pathways that underlie "memory" and elaboration of new functional links" to quote Bernard Katz [1966]. What might be the reason for such deep interest manifested by "pure" neuroscientists to the problem of memory, which for the prolonged periods of time had been the "promised land" for only speculative psychology and philosophy?

We think that main reason was current success in behavioral science which resulted in the following statement of fundamental importance:

behavior of higher vertebrate animals could not be viewed as a simple "chaining" of individually acquired conditioned reflexes (basic principle of "behaviorist" psychology and Pavlov's views of so called "physiology of higher nervous activity") but is determined by some "internal representations" of stimuli perceived earlier in specific surrounding environment. Retrospectively two main opponents of behaviorist concept of animal behavior may be mentioned – psychologist Edward Tolman in USA and neurophysiologist Ivane Beritashvili ("Beritoff" or "Beritov" in Russian, and in Western languages) in Georgia. E. Tolman's ideas about determining the role of the "internal representations of earlier perceived events" in the organization of mammalian behavior are well known all over the world and by this reason we will not accentuate attention on them [Tolman,1932].

Still in 1932 Beritashvili proposed new original concept about animal behavior [Beritov, 1932]. In his view animal behavior in quite new situation is determined not by conditioned reflexes but by "images" of spatial locations of vitally important stimuli. Numerous and diverse experiments convincingly showed that animals (dogs) in the new situation after the first perception of vital stimuli (food or some damaging stimulus) manifest not the chaotic, but directed, expedient behavior. This expediency of behavior is caused by the regulating role of the images of the external situation, which are created with the perception of appropriate objects in it and which are reproduced in the same situation through one or other time interval.

From our point of view extremely important in the concept of I. Beritashvili is not simply the assertion about the presence of concrete images in animals, but the idea about the active role of these reproduced images in the determination of adaptive behavior; in other words, according to I. Beritashvili "images" appear not as the epiphenomenon, but as the active main regulator of the behavior of animals in the new situations.

From the other side, theoretical studies of the forms of behavior in the self-organizing systems, based on the cybernetic ideology showed that as the conceptual construction " the chain of conditioned responses" principally cannot explain the experimentally verified forming of expedient behavior in so-called " emergency situations]", i.e., when system must in a short time be adapted either to the completely new, suddenly prevailing situation or to an abrupt change in the old situation; it turned out that in all these cases system can organize adaptive activity only when it acts not according to the principle of algorithms (" the chains of conditioned reflexes but according to the principle of heuristics (Bernstein N. A. 1966; Bongard M. 1967, Minski M, ; Miller J., Gelenter E., Pribram K. H.,). Biological example of such " emergency situations" can serve the possible encounter of animal with the enemy, when in the very short time it is necessary to make a decision, either about the active avoidance (situation victim - predator), or about the active attack (situation predator - victim) and so forth.

In all such cases of "the chain of conditioned reflexes", manufactured in the old situation on the basis of the multiple repetition of the specific behavioral responses under the constant conditions, cannot ensure sufficiently rapid adaptation to situation, i.e. in other words, living organism, acting in the new situation according to the principle of elaboration of the conditioned reflex would perish from the hunger, the thirst, the predator, the chaotic action, or simply inaction faster than would be manufactured corresponding optimum of "the chain of conditioned reflexes", that requires repeated reinforcements of surrounding stimuli with the rewards and subsequent production of the reflexes of the second and higher orders. But in the real life organism frequently is adapted to the new situation quite rapidly "in one trial" , and this is ensured according to experimental data and theoretical ideas of [i]. [Beritashvili] by the precisely descriptive psycho-nervous activity.

This functional role of the image in the behavior is manifested in its motor activity, because of which animal is not simply adapted to the environment (passive balancing with the medium by I. Pavlov), but actively overcomes its obstacles for achievement of the specific goal of behavioral act represented in the images [Beritoff,1965]; see also the general concept of the "physiology of activity" in the work of Bernstein N. A. [1966]. Its "automation" and conditioned-reflex chaining occurs only with the multiple repetition of this behavioral pattern under the constant stable conditions of the surrounding situation and it begins to proceed as a "chain conditioned reflex" [Beritoff, 1965].

The ideas, similar to the above ideas of I. S. Beritashvili were in many respects developed by a Soviet physiologist P. K. Anokhin, according to whom the leading factors, which determine behavior are the "afferent synthesis" and the "acceptor of action" , into which is packed almost the same content, as in the term of "image" [Anokhin, 1979]. Close ideas were advanced by another Soviet physiologist N. A. Bernstein [1966], who on the basis of the detailed experimental and mathematical analysis of the motor activity of animals came to the conclusion that the forming of reaction occurs not according to the principle of "stimulus - response", but according to the continuous corrections according to the principle of feedback. But in that case in the central nervous system unavoidably must exist in the coded form the prior anticipation of the required final result of reaction, i.e. must exist " the model of the required future", since only comparison of the arriving afferent information about the existing situation with that provided by "the model of the required future" can in fact ensure the expediency of motor response [Bernstein, 1966]. In this connection it is interesting to note that this continuous afferent correction according to the principle of feedback is necessary precisely for the realization of behavior, adjusted by images, whereas the automated conditioned-reflex behavior it does not require [Beritov, 1969]. From the other side, the founders of cybernetics, examining behavioral patterns, characteristic for the complexly organized systems, came to the conclusion that one of the "highest" types of behavior, characteristic of animals appears to be "teleological" - the behavior, by which they mean the behavior, allotted by activity, by endeavor and by adjustability according to the principle of feedback [Rosenblueth, Wiener, Bigelow, 1943]. Finally, the "images" of objects moving in the space, which lie at the basis of the capability of animals for the prediction of the future sections of the trajectories of motion, intensively were studied by L.Krushinskiy, who isolated this type of behavior into the group of the so-called " extrapolation reflexes"; it is remarkable, that this type of behavior is achieved " from one trial" that is after the first test, but it is not manufactured according to the principle of the conditioned reflex [Krushinskiy, 1966].

Thus all contemporary data and theoretical concepts of the behavioral science with the obviousness indicate the important role of the "images" of the external world, preserved in the memory, in the organization and regulation of the adaptive behavior of "higher" animals.

Being the students of I. S. Beritashvili, we, naturally paid primary attention to the accentuation of his ideas about the role of "image-driven memory" in the organization of the behavior of animals. Contemporary studies of the visuo-spatial recognition memory in mammals show the possibility of the experimental separation of the engram of " the location of the stimulus in space " ("image of the stimulus spatial location" according to I. S. Beritashvili) to its constituting components - "stimulus" *per se*, irrespective of its spatial location (thus "nonspatial visual memory") and "place" to be remembered *per se,* irrespective of its "content" (thus "spatial visual memory").

II. Nonspatial Visual Recognition Memory in Cats

As it is well known in experimental psychology the evaluation of memories and their characteristics in people are produced by two basic methods: by the method of “free recall" and by the method of "recognition". As far as we know, for evaluating the memory in animals thus far effectively is used only the second of those methods. Since this part is dedicated particularly to memory in subhuman animals, speech subsequently will concern in essence precisely only the recognition memory, with small exceptions. At the same time we were deeply impressed by the success obtained by neuropsychologists in the USA in the study of recognition memory in animals (mainly in monkeys and rats), especially by the work of M. Mishkin’s team at Bethesda laboratory of Neuropsychology.

The selection of cats as experimental animals for investigating the memory of recognition was caused in essence by the following bases: 1. As far as we know the recognition memory was investigated mainly in the primates and rats; therefore it appeared to be interesting to establish its presence or absence in the cats as representatives of carnivores, in particular if we consider the fact that the cat, until now, remains classical object for the neurophysiological studies.

Delayed matching- and nonmatching-to-sample (DMS and DNMS) tasks have been widely used to study one-trial object recognition in monkeys [Gaffan, 1974; Mishkin, Delacour, 1975; Mishkin,1982] and to elucidate the cerebral mechanisms that mediate this type of cognitive visual memory [Mishkin,Murray, 1994; Squire, 1992]. Similar studies have been carried out in rodents [for review see Steckler, Drinkenburg, Sahgal, Aggleton, 1998] and other mammals, such as dogs [Callahan, Ikeda-Douglas, Head, Cotman, Milgram, 2000]. But to our knowledge there are no such reports on cats, although some researchers have shown that cats have good memory for spatial locations as measured by their performance on the classical spatial delayed response task [Beritashvili, 1971; Fletcher, 1965; Konorski 1967; Warren, Warren, Akert, 1972]]. The purpose of the present part is twofold – one is to uncover the possibility of normal cats to solve different versions of nonspatial delayed matching- and nonmatching-to-sample tasks and the other – to uncover the possible brain regions perhaps responsible for the nonspatial visual memory in cats.

We believe that the most satisfactory neurobehavioral studies addressing interactions between the brain and behavior have until recently been performed in primates. On the other hand, the preferred study system for purely neurophysiologic studies has to date been the cat. Both fundamental and comparative studies with a wide evolutionary perspective need these two approaches to be integrated. The major point here is that the whole arsenal of state-of-the-art neurophysiologic methods can be used in cats for identification of the basic neuronal mechanisms of recognition memory. With this purpose we elected to study in cats some of the forms of cognitive memory which have been well studied in monkeys. In particular, this relates to visual recognition memory in conditions of one-trial visual recognition, which has previously been studied in the USA [Mishkin, Delacour, 1965; Mishkin, Prockop, Rosvold, 1962] using original methods, i. e., image-based delayed selection. Results of such study already have been presented by us earlier. As regards the cerebral mechanisms of this type of visual memory, Murray, Bachevalier, Mishkin [1989] and Meunier, Bachevallier, Mishkin, Murray [1977] showed that the perirhinal and entorhinal cortical areas in primates together (i. e. the "rhinal area") play the key role in the organization of this type of memory. Is the link between this type of memory and the rhinal area a characteristic of primates or is it more widespread among mammals? As it has been demonstrated by other investigators in dogs [Callahan, Ikeda-Douglas, Head, Cotman, Milgram, 2000] and by us in cats [Okujava, Natishvili, Mishkin, Gurashvili, Chpashvili, Bagashvili, Andronikashvili, Kvernadze, 2005] as representatives of carnivores they can solve the visual nonspatial recognition memory tasks. On the other hand studies in rats have demonstrated not only the existence of visual nonspatial memory, but also the involvement of the rhinal cortical areas in performing the corresponding tasks [Mumby, Pinel, 1997].

In the context of neuropsychological studies of memory, our specific intent was to use the "delayed matching-to-sample" and "delayed nonmatching-to-sample" tests to determine the contribution made by these cortical areas in cats to one-trial non-spatial visual recognition memory.

Materials and Methods

Experimental subjects

Fifteen experimentally naïve adult normal cats of both sexes (eight male, seven female) weighing 3-4.7 kg were used in this study. The animals were housed in individual cages (1.5 x 1.0 x 1.0 m) in which they had free access to water. Food was given once daily, 20 h before testing. Experimental sessions were conducted 5 days per week. The care and use of the animals complied with Georgian regulations, with Guidelines prepared by the Ethics Committee of the Institutional Animal Care and Use Committee of the Research Center for Experimental Neurology, and with the National Institutes of Health Guide for the Care and Use of Laboratory Animals.

Five different tasks were used, two requiring manipulatory responses, and three – locomotor responses

TASK I: M-DMS (MANIPULATORY VARIANT OF DMS). Four cats, two of them males, were trained on this task in modified Wisconsin General Testing Apparatus (WGTA).

Apparatus

The Wisconsin General Testing Apparatus (WGTA) was adapted for use with cats so that they could use their forelimbs to displace objects and retrieve food. The apparatus as described in detail by Okujava, Natishvili, Mishkin, Gurashvili, Chpashvili, Bagashvili, Andronikashvili, Kvernadze [2005] consisted of two main parts (Figure 1): A cat start-cage (55cm x 65cm x 60 cm) placed on a table inside a darkened, sound-shielded room; and a test tray containing three identical food wells, each a round glass jar (25 mm deep and 73 mm in diameter), on which different objects (stimuli) may be placed. The stimuli consisted of an array of 600 junk objects, which differed from each other in size, form, texture, and color (the latter providing mainly brightness cues for cats).

Procedures

During *preliminary training*, cats were shaped behaviorally to displace cardboard covers placed over the three food wells to obtain rewards (each a small piece of boiled meat, 0.5 cm^3) hidden in the wells. They were then trained in the same way to displace one of three pretraining objects, which were presented singly in random order over one of the three food wells. Finally, the cats were given 20 pseudotrials to familiarize them with the structure of the task: one of the three pretraining objects was presented as the "sample" object over the baited central well; 10 s later the two other objects were presented over the lateral wells, both or neither of which were baited, in random order. The cat was allowed to displace only one of the two "test" objects. The pseudotrials were separated by 30 s intervals. During the 10 s delay intervals and the 30 s intertrial intervals, an opaque screen separated the cat from the test tray. This preliminary training was completed in 7-12 days. *Formal testing* was then begun, using trial-unique objects.

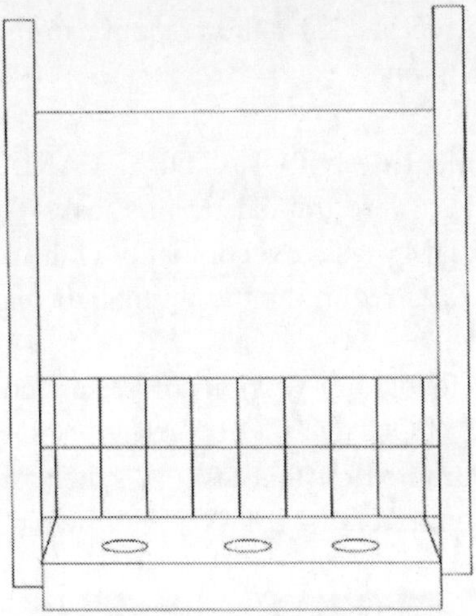

Figure 1. Schematic sketch of the WGTA-type apparatus used for tasks I and II. The side of the cat cage facing the test tray consists of metal bars. The tray, located 15 cm in front of the bars, contains three food wells 15 cm apart (center to center). The opaque screen between the cage and the tray is shown in the raised position.

TASK I: M-DMS (MANIPULATORY VARIANT OF DMS) task: each trial consisted of two parts, a sample presentation followed by a choice test (Figure 2).

After the animal displaced the sample object from the central well and retrieved the reward (no other object was on the test tray), the opaque screen was lowered for delay intervals of 5 and 10 s in pseudorandom order. The screen was then raised revealing the sample object again together with a novel object, each covering one of two lateral wells, and the cat was allowed to choose. A new pair of objects was used on every trial, and the left-right positions of the sample and novel objects on the choice test varied pseudorandomly. In the choice tests of this task, the sample object was always baited, requiring the animal to learn the rule of delayed matching-to-sample. Twenty such trials, separated by 30 s intertrial intervals, were presented daily until the animal achieved the criterion score of 80 correct trials in 100 across five consecutive sessions. The time limit for the behavioral response was set initially to 10 sec, and to 5 sec at final stages of training; withholding the response beyond that limit was scored as error. The response of the animal toward the central food well was scored as an error. There was no correction for errors, *e.g.* an animal after making an incorrect response to central well was not allowed to correct itself by the response to the side well in the same trial, nor in the following one.

TASK II: M-DNMS (MANIPULATORY VARIANT OF DNMS). Four cats, two of them males, were trained on this task. The procedures were the same as those used for DMS task, except that now the novel object was always baited on the choice test, requiring the animal to learn the rule of delayed nonmatching-to-sample.

Let us turn now to locomotor versions of these recognition memory tasks.

TASK III: L-DMS (LOCOMOTOR VARIANT OF DMS). Four cats, one of them a male, were trained on this task. A schematic representation of the Nencki-type testing room [Konorski, 1967] is shown in Figure 3.

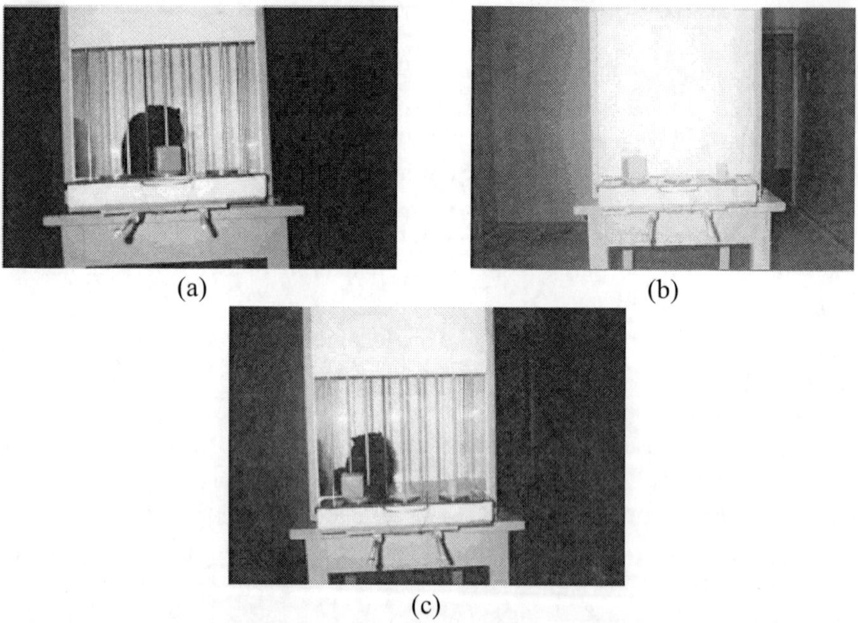

(a) (b) (c)

Figure 2. Successive stages of the "delayed matching to-sample" (DMS) test procedure is shown below the photo of the apparatus with cat. A – sample presentation phase; B – delay phase; C – choice phase.

The cat cage (labeled as "St.c" in Figure 3), measuring 45 cm x 45 cm x 50 cm was equipped with an opaque screen. To facilitate perception of the stimuli from the cat cage, only the stimulus array's larger objects (15-20 cm high and 3-5 cm in diameter) were used in this and following tasks.

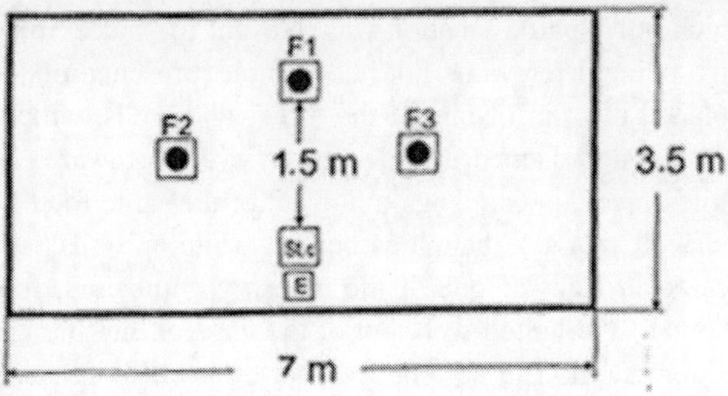

Figure 3. Schematic representation of the Nencki-type testing room used for Tasks III-V. Abbreviations: (St.c) start-cage; (F1, F2, F3) three identical food wells; (E) location of the experimenter. Arrow inside the figure (calibration bar) indicates the approximate distance between the syart-cage and the food wells.

The three stages of preliminary training were similar to those given in WGTA, except that the animals were shaped behaviorally to run to the food wells to retrieve rewards, and then to return to the cage and reenter it. The preliminary training was completed on average in 2 days.

Both for the pseudotrials and during formal testing, objects were placed in front of the baited central food well (F1 in Figure 3) for the sample presentation, and in front of the lateral wells (F2 and F3 in Figure 3) for the choice test. When the animal returned to the cage from the central food well after the sample presentation, the opaque screen was lowered for the delay period (5 s and 10 s in random order, as before), after which it was raised so that the animal could choose between the lateral food wells marked by sample and novel objects. On this task, as in Task I, the food well marked by the sample object was always baited on the choice tests, requiring the animal to learn the rule of matching-to-sample. In all other respects also, the procedures for this task were the same as those used for Task I.

TASK IV: L-DNMS (LOCOMOTOR VARIANT OF DNMS). Three cats, all males, were trained on task IV. All procedures were the same as those used in Task III, except that on the choice tests the novel object was always baited, requiring the animal to learn the rule of nonmatching-to-sample.

TASK V: L-DR (LOCOMOTOR VERSION OF SPATIAL DELAYED RESPONSE). The four cats that had been trained on Task III were transferred to Task V. This task was adapted from the "place trials" of the object-location task described by Parkinson and coauthors [Parkinson, Murray, Mishkin,

1988], in which only spatial location was relevant for successful performance, object quality being irrelevant. For the sample presentation, a trial-unique object was placed in front of one of the three food wells, which were baited, and which the cat approached in order to retrieve the reward. For the choice test, the sample object appeared again in front of the same food well as before, and this food well was also baited as before, while an object identical to the sample appeared in front of one of the other two, unbaited, food wells. The spatial locations of the baited well and of the covered but unbaited well were selected pseudorandomly. In all other respects, this task was the same as Task III (L-DMS).

TESTS WITH EXTENDED DELAYS: TASKS I-IV. Immediately after completing training on the 10 s delay, each cat was retrained to criterion on the same task it has received before, but at a 30 s delay. Two weeks later each cat was tested for retention at the 10 s and 30 s delays, retrained to criterion if necessary, and then trained to criterion at both 5 min and 10 min delays. These tests with extended delays were presented in 20 trial daily sessions in pseudorandom order. Statistical assessments of the behavioral data were performed on the number of trials and errors preceding criterion at each training or retraining stage by Mann-Whitney U test [Hollander, Wolfe, 1973].

For the second subsection of this part only the M-DMS task was used for control and operated cats (bilateral lesions of the "rhinal region"); as general behavioral procedures of M-DMS testing were already outlined above, here we will consider only the surgical methods employed for this subsection. For this reason we performed another study of nonspatial visual recognition memory in new group of 12 adult cats of both genders (seven male, five female), weighing 3-3.5 kg, not previously used in experimental studies. Between experiments, animals were kept in the animal house in individual cages (1.5 x 1.0 x 1.0 m) in which they had free access to water. Food was given once daily, 20 h before testing. Experimental sessions were conducted 5 days per week. Same modified WGTA apparatus was used in this study as already described in preceding section of the Chapter. M-DMS task described previously was used for one-trial nonspatial visual recognition memory testing with all parametric data (trials per experimental day, criterion of performance, delay intervals, intertrial intervals and so on) just identical to previously described.

Surgery

After completion of preliminary training animals were divided into three independent groups each of four cats. In one group animals received lesions to the rhinal (i. e., perirhinal + entorhinal) cortical area (group R, n=4). In the next group, cats underwent all surgical manipulations except for lesioning of the structures of interest (sham-operated animals, SOP, n=4). The last group consisted of non-operated control animals (group C, n=4). All surgery was performed in strictly aseptic conditions on animals under deep Nembutal (40-60 mg/kg) anesthesia. Electrolytic lesions of cortical areas were produced by passing an anodal current of 2-5 mA through an electrode of stainless steel, diameter < 0.2 mm, insulated with epoxy resin throughout its length except the final 0.5 mm at the tip. The coordinates for electrolytic rhinal lesions were selected using a stereotaxic atlas of the cat brain [Reinoso Suarez, 1961]. Cats in the SOP group underwent the same surgical procedures but no current was passed through the electrode. Coagulation of the whole rhinal region was produced by an average of 10-15 insertions of the electrode through small openings (diameter 0.5-1 mm) made in the cat's skull with a drill. Preliminary experiments yielded histological evidence showing that the diameter of the electrolytic lesion of cortical tissue produced by one insertion of the electrode and application of the current with the parameters noted above averaged 3-5 mm (Figure 4, insert).

After all surgical procedures were complete, the margins of the surgical wound were sutured. The post-operative recovery period lasted 7-10 days, after which testing was started. The borders of the perirhinal and entorhinal cortex are shown in Figure 5. By analogy with results obtained previously in other mammals [Krettek, Price, 1977; Witter, Groenewegen, Lopes da Silva, Lohman, 1989], the perirhinal region in cats is presumptively located along the posterior rhinal sulcus, occupying the base and lateral wall of this sulcus (Figure 5, A,B).

Its anterior margin lies caudal to the sylvian sulcus (s), where the cortex in the depth of this sulcus is adjacent to the temporal cortex of the posterior composite sulcus. The posterior margin of the perirhinal area is located at the caudal margin of the posterior rhinal sulcus, where the latter splits into the medial-running pararecurrent sulcus [Kreiner, 1971]. The entorhinal cortex (EC) occupies the posterior part of the piriform cortex, which is the ventral part of the limbic lobe. The anterior margin of the entorhinal cortex is adjacent to the prepiriform cortex (PPC) and the cortical amygdaloid nucleus (Co). Laterally, the entorhinal cortex runs to the medial wall of the rhinal sulcus (rhp), where it is adjacent to field 35 of the perirhinal cortex, as medially it reaches the parasubicular area of the hippocampal formation. The posterior margin of the entorhinal area consists of the retrosplenial sulcus, separating the entorhinal cortex from the post- and retrosplenial cortical areas located on the medial surface of the hemisphere.

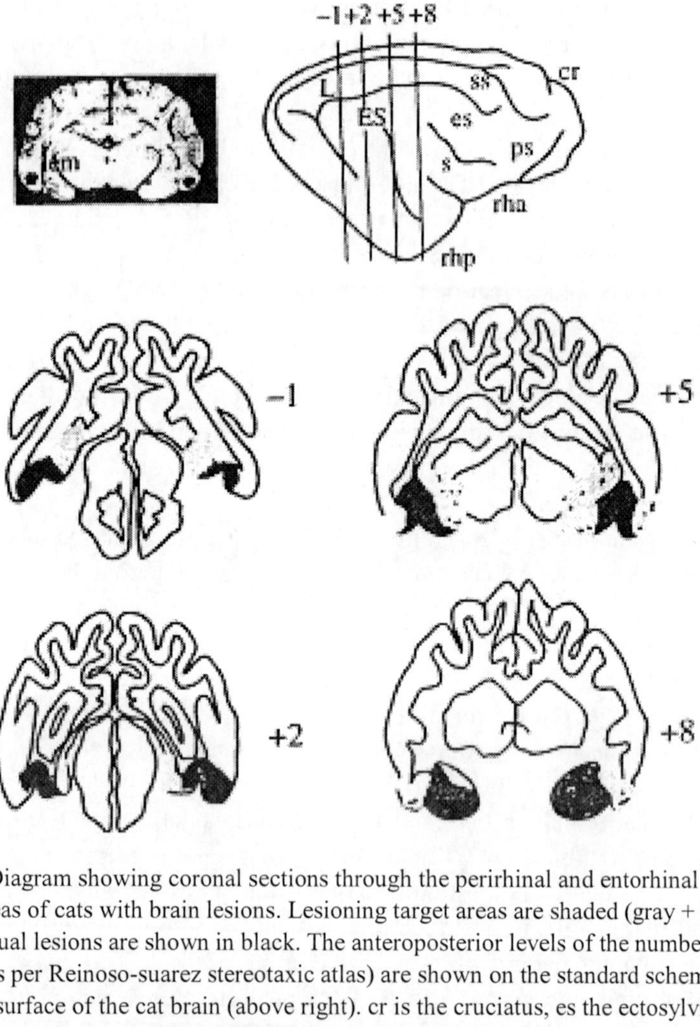

Figure 4. Diagram showing coronal sections through the perirhinal and entorhinal cortical areas of cats with brain lesions. Lesioning target areas are shaded (gray + black areas). Actual lesions are shown in black. The anteroposterior levels of the numbered sections (as per Reinoso-suarez stereotaxic atlas) are shown on the standard scheme of the lateral surface of the cat brain (above right). cr is the cruciatus, es the ectosylvian, L the lateral, ps the presylvian, rha the anterior rhinal, rhp the posterior rhinal, s the sylvian and ss the suprasylvian sulcus. The insert shows a transverse section at coronal level +5, showing the approximate size of the typical electrolytic lesion applied through a single electrode.

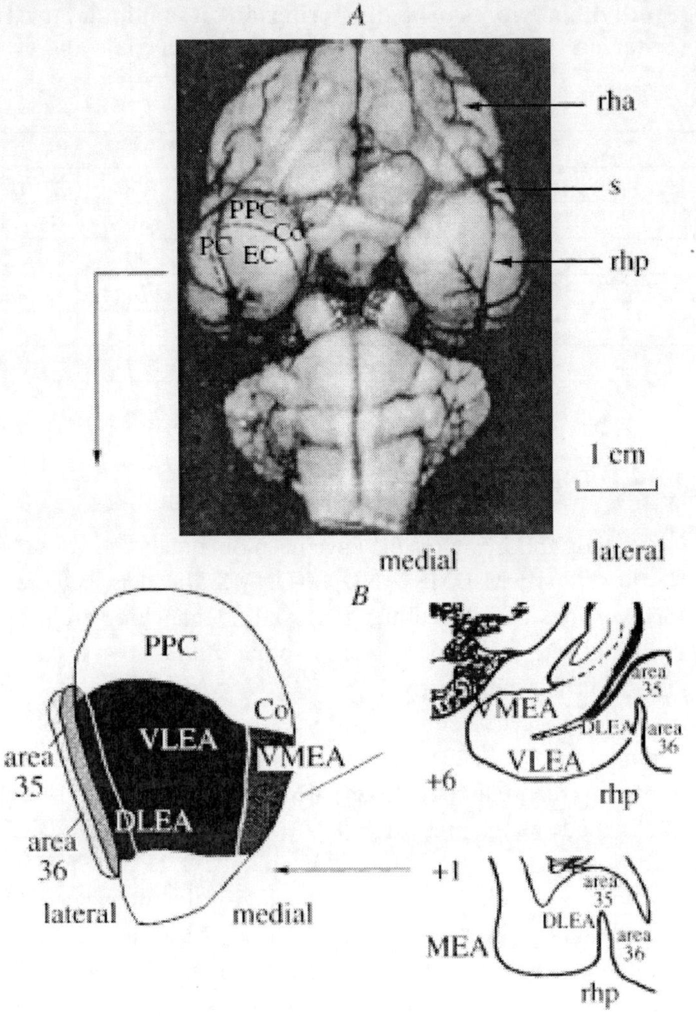

Figure 5. Photograph of the ventral view of a cat brain showing structures of the piriform cortex of the limbic lobe destined to be lesioned. (B) Enlarged diagram of the piriform cortex in the right hemisphere (left side of figure) shows cytoarchitectonic subdivisions of the perirhinal and entorhinal cortical areas. Arrows and diagrams indicate the levels of coronal sections through to be lesioned regions. Abbreviations: 35, 36, Brodmann's subdivisions of piriform cortex; Co, cortical amygdaloid nucleus; DLEA, dorsolateral entorhinal area; EC, entorhinal cortex; MEA, medial entorhinal area; PPC, prepiriform cortex; rha, anterior rhinal sulcus; rhp, posterior rhinal sulcus; s, sylvian sulcus; rs, VLEA, ventrolateral entorhinal area; VMEA, ventromedial entorhinal area.

Table 1. Individual scores preceding criterion on manipulatory Tasks I and II (criterion sessions were not included into the trials and errors to criterion indices)

Task I: M-DMS			Task II: M-DNMS		
Cats	Trials	Errors	Cats	Trials	Errors
Alpha	190	66	Anna	180	91
Beta	260	92	George	200	75
Gamma	350	116	Strong	170	50
Delta	220	75	Nancy	220	84
	240	84	MEDIAN	190	80

Results and Discussion

In present study following results have been obtained:

TASKS I AND II: M-DMS AND M-DNMS.The data gathered on the manipulatory versions of delayed matching and nonmatching are presented in Table 1. The eight cats trained on one or the other of these two tasks learned in an average of 224 trials and 81 errors.

Although the group trained on DNMS learned slightly more quickly then the group trained on DMS, the difference was not significant. These results differ from those reported in monkeys [Mishkin, Delacour, 1975], which showed a distinct advantage in learning DNMS as compared with DMS. It was concluded from the latter set findings that monkeys have a preference for novelty, a preference that hastened their learning of the nonmatching rule. On the basis of the findings in the present study, it appears that cats either do not have this preference for novelty or do not exhibit it to the same degree as monkeys do.

These conclusions regarding a species difference in preference for novelty and the effects of this preference on learning in cats and monkeys are supported by a direct comparison between the scores of the two species. Comparing our cats with monkeys from Mishkin and Delacour [1975] study it must be noted that in these two cases different learning criteria were used (80% in 100 consecutive trials for cats vs. 90% in 40 consecutive trials for monkeys). From the statistical point of view such comparison at first may seem not to be meaningful. But let us consider the data from Mishkin and Delacour paper: it is evident that their monkeys in M-DMS task ("E-M" in their notation), before reaching 90% criterion, reached criterion of 80% correct responses within 40 trials (median for trials to reach this criterion - 320), after which their learning curve raised monotonously to higher levels of performance (90% correct responses in 40 trials); thus it may be assumed that their monkeys in criterial trials in M-DMS task performed at least as efficiently as our cats in their 100 criterial trials. As to the M-DNMS task ("E-NM" in Mishkin and Delacour notation) monkeys reached criterion of 80% correct responses in at least 60 trials to criterion in spite of non-monotonous character of their learning curve [Mishkin Delacour, 1975; Figure 1]. It is worthwhile to mention that according to our results cats after having achieved 80% correct criterion in 100 trials performed subsequent lengthened delay tasks at least at the same (80%) proficiency level in both tasks (Table IV). From these statements one may conclude that it is meaningful to compare our cats with monkeys from Mishkin and Delacour work mentioned above. The eight monkeys in the study by Mishkin and Delacour [1975] learned one or the other version of the two tasks in an average of 225 trials and 85 errors, an overall mean score remarkably similar to that of the eight cats in the present study. However, the monkeys' scores were distributed very differently between the two tasks: DMS, 360 trials and 150 errors; DNMS, 90 trials and 21 errors. As a result, although the cats learned the nonmatching rule significantly more slowly than the monkeys did (two- tailed Mann-Whitney $\boldsymbol{U_{4,4}=0}$, P=0.03, for both trials and errors), they tended to learn the matching rule slightly more quickly than the monkeys, though the difference in this case is not significant.

TASKS III AND IV: L-DMS AND L-DNMS. The results obtained in the Nencki-type testing room are presented in Table 2.

Compared with the cats trained in the WGTA, those trained on the locomotor versions learned relatively slowly, with some animals even failing to reach criterion within the limit of 1 000 trials. The cats in this situation exhibited a strong tendency to return to the central food well on the choice test, suggesting that visuospatial strategies, which are well developed in cats [Warren, 1965; Warren, Warren, Akert, 1972], had interfered with their learning of the object matching and nonmatching rules. To examine how readily cats might learn a locomotor task in which memory for spatial cues was relevant and memory for object cues was irrelevant, we transferred the animals that had been trained on task III to Task V.

TASK V: L-DR. As shown in Table III, all four cats, including the two that had previously failed to learn the locomotor version of delayed object matching, learned the locomotor version of spatial delayed response within just four sessions (i.e., 80 trials).

Table 2. Individual scores preceding criterion on locomotor Tasks III and IV (Criterion sessions were not included into the trials and errors to criterion indices)

L-DMS			L-DNMS		
Cats	Trials	Errors	Cats	Trials	Errors
Tetra	1060 (F)	520	#1	510	245
Shava	640	255	#2	725	350
Natsara	900	390	#3	1000 (F)	470
Tsotskhala	1040 (F)	435			

F denotes failure to achieve criterion within 1000 trials. Although two cats (Tetra and Tsotskhala) achieved criterion in additional 6o and 40 trials, respectively, while on the third cat (Natsara) training was disconnected after 1000 trials.

Table 3. Individual scores preceding criterion on the locomotor spatial memory task, Task V: L-DR (criterion sessions were not included into the trials and errors to criterion indices)

Cats	Trials	Errors
Tetra	60	25
Shava	40	15
Natsara	80	45
Tsotskhala	80	38

Table 4. Median errors to criterion for initial learning of Tasks I-IV at 10 s and 30 s, and for relearning after a two-week rest at both 10 s and 30 s and at extended delays of 5 min and 10 min (criterion sessions were not included into the trials and errors to criterion indices)

MEDIAN ERRORS						
Tasks	Learning		Relearning		New learning	
Delays	10 s	30 s	10 s	30 s	5 min	10 min
M-DMS	84	90	0	2	6	2
M-DNMS	80	86	0	0	4	0
L-DMS	413	310	0	4	6	0
L-DNMS	350	325	0	0	4	0

The results support the suggestion that the cats' strong tendency to use visuospatial strategies in the large testing room interfered with their learning of the object matching and nonmatching rules in that situation.

PERFORMANCE AT EXTENDED DELAYS: TASKS I-IV. As shown in Table 4, those cats that had mastered object matching and nonmatching up to delays of 10 s committed about the same number of errors as before to master same task at a delay of 30 s.

Once they had attained the criterion score at 30 s, however, the cats re-attained this criterion almost immediately both when tested for retention after the two-week rest period and when delays were subsequently increased to 5 min and 10 min.

Previous comparisons of the visual learning and memory abilities of cats and monkeys were based mainly on the use of the learning-set paradigm [Passingham, 1981; Warren, 1965], and these comparisons indicated inferior performance of cats on trial 2, a measure of one-trial memory. Based on the evidence gathered on the tasks used in the present experiment, however, the one-trial memory of cats seems to be quite comparable to that of monkeys, in terms of both overall speed of rule learning, at least in the WGTA, as well as level of retention over long intervals [Mishkin, 1982]. The results suggest that delayed matching- and nonmatching-to-sample are useful measures for investigating the neural basis of "long-term" visual memory in cats. During comparison of cats' performance on the visual recognition task with the performance of monkeys the following distinction is evident: cats perform the locomotor version of the recognition tasks poorly, due to their reliance on visuospatial strategies directed to find food in a particular place visited a bit earlier. Perhaps this distinguishes them, as a group, from monkeys, although this speculative point needs special comparative study. It might be supposed that this difference is caused to a substantial degree by cats' inherent tendencies to exhibit the predatory behavioral patterns directed mostly by visual checks of the spatial locations of relevant objects [Warren, 1965]. These tendencies, of course, should be present in primates too, but to a much lesser extent, because monkeys' manipulatory activity is more developed.

Altogether, it appears that cats, like monkeys, have a highly developed "long-term" visual recognition memory ability. At the same time our already reported results indicate the urgent need to investigate in cats the visuospatial recognition memory *per se,* although first of all, to be logically complete we heed now to report our data concerning possible brain mechanisms responsible to mediation of visual nonspatial memory in cats. Now, our following research step consisted in an attempt to uncover possible brain region sub serving nonspatial visual recognition in cats.

Morphological Monitoring

After behavioral testing, cats were overdosed with barbiturate, perfused via the heart with normal physiological saline followed by aldehyde fixatives. The brain was removed and photographed and ~~50-m~~ cryomicrotome sections were cut; each fifth section was stained with thionine and was covered with a cover slip. Sections were examined histologically. The areas of cell loss and gliosis, along with the extent of the lesions, were superimposed on diagrams of intact coronal sections. Lesioned areas of structures were measured in square millimeters on drawings of each section. Lesion margins were determined using a simple digital system running on computer and subsequent manual calculation of lesion size. The volumes of lesioned tissue in the areas of interest were expressed as percentages of the normal volumes of the areas. The results for each case are presented in Table 5. Lesions to the rhinal area, regarded as combined lesions to the perirhinal and entorhinal areas, were incomplete (mean 75%, range 54-88% for the entorhinal cortex, mean 80%, range 64-89% for the perirhinal cortex). All animals showed some unplanned bilateral lesioning of the parasubiculum and presubiculum (mean 48%, range 33-56%; Table 5, Figure 4), though unexpected lesioning of other areas outside the rhinal cortex was insignificant.

Table 5. Extent of damage to medial temporal lobe structures in percents to total volume of corresponding structures

Volume, mm^3	HF (406)			Para/Presub (70)			ERh (213)			PRh (233)			Compo (406)		
Cats	Lt	Rt	C	Lt	Rt	C	Lt	Rt	C	Lt	Rt	C	Lt	Rt	C
R. No. 1	0	2	1	61	51	56	98	83	91	56	72	64	6	3	5
R. No. 2	0	0	0	43	57	50	75	83	79	70	90	80	26	20	23
R. No. 3	14	12	13	32	34	33	95	44	70	90	88	89	9	21	25
R. No. 4	9	11	10	51	57	54	85	40	63	92	82	87	29	22	26
Mean	6	6	6	47	50	48	88	63	76	77	83	80	18	17	20

HF, Hippocampal formation; Para/Presub, parasubiculum plus presubiculum; ERh, entorhinal cortex; PRh, perirhinal cortex; Compo, area composita posterior according to Kreiner (1971); Lt, left; Rt, right;

Table 6. Individual scores preceding criterion (criterion sessions were not included into the errors to criterion indices) in DMS Task tested under "short" (5 sec, 10 sec) delay conditions in normal, sham operated (control animals groups) and operated rhinal cats.(criterion – no more than 20 errors in 100 consecutive trials

GROUPS:	ANIMALS	DELAYS	
		5 sec	10 sec
NORMAL CONTROL(n=4)	# 1	66	55
	# 2	92	100
	# 3	116	90
	# 4	75	48
SHAM OPERATED CONTROL (n=4)	# 1	102	62
	# 2	69	70
	# 3	75	102
	# 4	100	65
RHINAL OPERATED EXPERIMENTAL (n=4)	# 1	90	120
	# 2	87	110
	# 3	100	130
	# 4	105	150

This included minor bilateral lesioning of the hippocampus itself (mean 6%, range 0-13%) and the area composita (mean 20%, range 5-26%) according to Kreiner's notation (1971), this being unilateral in the latter case – cat R. No. 4 (Table **5**, Figure 6).

Consequences of Lesioning of the Rhinal Area

Statistical analysis of the results of the behavioral experiments was performed using a computer running the statistical packages Statistica and SPSS 10 (Statsoft Incx., USA). Analysis of the data presented in Table 6 shows that the results obtained from performance of the "delayed matching-to-sample" test by intact and SOP cats could be combined into a single group, termed "control animals".

Single-factor ANOVA demonstrated a statistically significant intergroup difference between control and experimental cats with rhinal lesions to the brain (F = 5.935; p = 0.009). On the other hand, the same analysis showed that the post hoc Sheffe's test for multiple comparisons identified a significant difference between cats with lesions to the rhinal area and animals of the control group with a 10-sec delay (p = 0.024) but not with a 5-sec delay (p = 0.999). This is entirely consistent with results obtained from nonparametric analysis and, furthermore, is additional evidence for the absence of any influence of lesions to the rhinal area on the initial "delayed matching-to-sample" training.

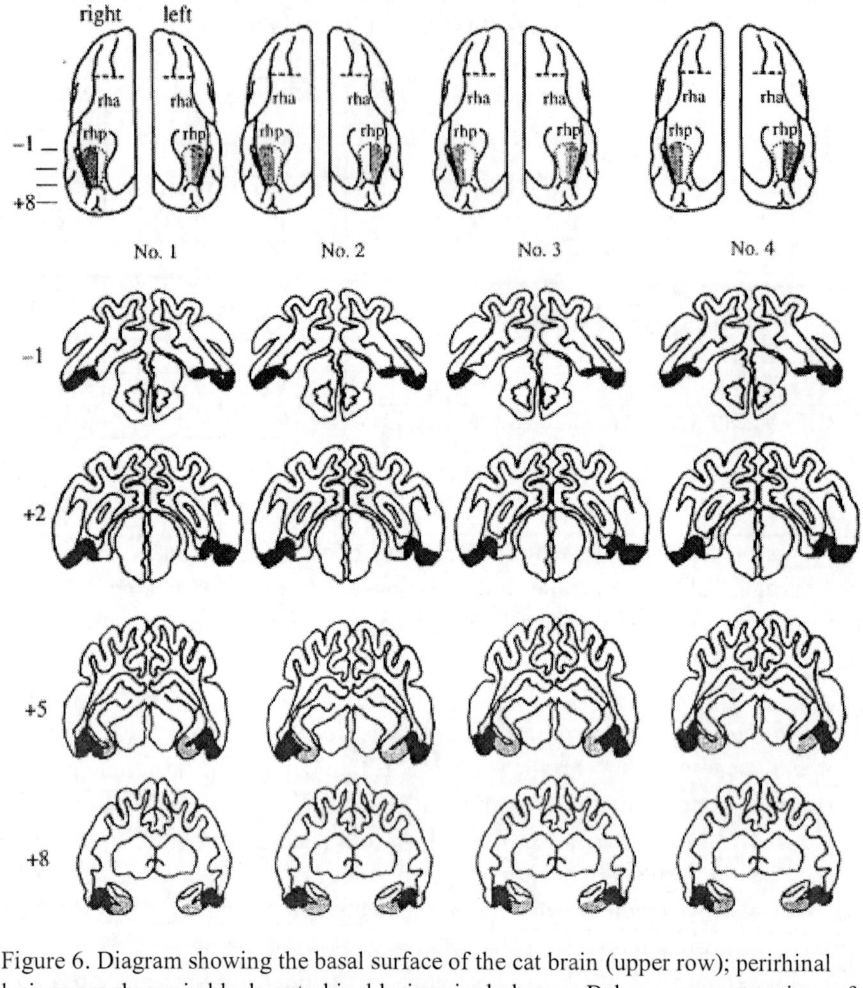

Figure 6. Diagram showing the basal surface of the cat brain (upper row); perirhinal lesions are shown in black, entorhinal lesions in dark grey. Below – reconstructions of perirhinal and entorhinal lesions (black) in four operated cats (R. No1, R. No 2, R. No. 3, and R. No. 4). Anteroposterior levels of numbered sections (as per Reinoso-Suarez stereotaxic atlas) are shown on the standard scheme of the ventral surface of the cat brain (upper diagram). Numbers indicate distances in millimeters from the interaural vertical plane. For further see captions of to Figures 4 and 5.

It seems appropriate to start the discussion of the data obtained here from the point of view of Mishkin's hypothesis of the hierarchical organization of memory for one-trial perception of objects in primates [Mishkin, Suzuki, Gadian, Vargha-Khadem, 1997]. According to this hypothesis, recognition memory for different sensory modalities (including vision) is organized hierarchically: the hippocampal system receives its main inputs from widely disseminated areas of the neocortex via two main projections - one from the perirhinal and one from the parahippocampal cortex. The first of these projections (the "ventral stream") arises from different sensory modalities important for analysis of stimulus quality, while the latter arises from the neocortical areas important for analysis of stimulus locations (the "dorsal stream"). We will not discuss dorsal stream here in this section as, according to the hypothesis, it is responsible for spatial memory, which was not addressed in the present part of this Chapter. The ventral stream, which is important for the organization of visual nonspatial memory, consists of "higher visual areas" (the inferior temporal cortex) and the perirhinal and entorhinal cortical areas; information is projected from these to the hippocampuses.

This raises the important question: is the ventral stream (or some analog!) present in carnivores? Verification of the existence of such a stream in cats clearly requires knowledge as to whether cats generally have an albeit functional analog (according to the results of Hodos and Campbell (1969) ideology) of the "inferior temporal cortex" if primates. Unfortunately, there are very few data on this important question and these data are contradictory: some investigators have shown the existence of such an analog in cats [Campbell, Jr., 1978; Hara, Cornwell, Warren, Webster, 1974; Natishvili, 1979], while others refute its existence [Gross, 1973]. In general, further investigations are needed to obtain a more strongly grounded interpretation of our data from this point of view. Our data suggest the conclusion that visual recognition in cats is mediated by the same cortical area as in monkeys. It should, however, be emphasized that our conclusion, like the analogous conclusion of American investigators with respect to visual recognition in primates [Murray, Bachevalier, Mishkin, 1989], is restricted only by short-term recognition memory. In our studies, animals of the experimental group (with lesions to the rhinal area) did not differ from controls on testing of recognition with delays shorter then 10 sec. This is evidently further support for the view that visual perception of an object and memory of it (even in the short-term version) depends on different brain structures in mammals. Overall, an important conclusion follows from the present data: visual recognition memory can be regarded as an "evolutionary invariant" in different mammal species. However these species have different ethological and ecological characteristics: it is sufficient to say that monkeys are diurnal, while cats are nocturnal.

To summarize obtained results we can make following main inferences: cats trained in the our modified WGTA learned the two tasks at about the same rate, on average, as that reported for monkeys. However, unlike monkeys, whose strong preference for novelty facilitates their learning of the nonmatching rule and retards their learning of the matching rule, the cats learned the two different rules at about the same rate, suggesting that cats do not share the monkey's strong preference for novelty. Iп contrast to their relatively rapid learning of the manipulatory versions of the two tasks, cats learned the locomotor versions only slowly or even failed to learn. Experimental analysis indicated that a major source of the cats' difficulty on these locomotor versions was interference from a strong tendency in the large testing room to use visuospatial strategies. Thus let us turn now to the study of visuospatial recognition memory in cats.

III. Spatial Visual Recognition Memory

Spatial memory, which is studied mainly in terns of delayed responses, is a phenomena which is widespread in different mammalian species [Beritashvili, 1971; Konorski, 1967; Nissen, Riesen, Nowlis, 1938; Olton 1977]. Among the various types of spatial memory, spatial recognition on one-trial perception of objects has only been studied relatively recently because of the development of original methods in Mishkin's laboratory in the USA [Parkinson, Murray, Mishkin, 1988]. Recent observations in primates have shown that the perirhinal cortex, as well as the entorhinal cortex and parahippocampal area, which make up the "medial temporal lobe", play a significant role in organizing the memory system [Mishkin, Suzuki, Gadian, Vargha-Khadem, 1997; Murray, 1992; Squire, Zola-Morgan, 1991; Zola-Morgan, Squire,Rasmus, 1994]. Lesions to the rhinal cortex in primates induce impairments to recognition memory, mainly in the visual and tactile modalities. Unlike early cytoarchitectonic evidence obtained by Brodmann [1909], the current view is that the perirhinal area extends more laterally and caudally along the posterior rhinal sulcus [Murray, Bussey, 1999]. The caudal perirhinal area in primates transitions to the parahippocampal (TF, TH) field, which is presently regarded as a component part of the functional memory system in primates. The analogous cortical areas in rats are fields 35 and 36 along the posterior part of the rhinal sulcus and the so called postrhinal area, located in the posterodorsal direction [Burwell, 2001]. Given the general nature of the cytoarchitectonic characteristics and the nature of the connections, this postrhinal cortex in rats is regarded as a homolog of the TF and TH fields of primates. In dogs, recent neuromorphological studies have shown that an area which is functionally equivalent to the parahippocampal in primates can be found in the most caudal part of the posterior rhinal sulcus [Woznicka, Cosmal 2003]. Considering the extensive similarity in the organization of the brain in felids and canines, we suggested that cats have an area functionally and topographically equivalent to the parahippocampal area of primates (Figure 7). On the basis of this suggestion, we report here studies of memory for spatial recognition on one-trial perception of objects in intact cats and cats with lesions to the possible analog of the posterior parahippocampal area in primates.

Materials and Methods

Experiments were performed on eight adult cats of both genders, weighing 3-4 kg. Experimental animals were placed in individual cages of size 1.5 x 1.0 x 1.0 m with free access to water. Food was provided daily at 20:00 prior to testing. Experimental sessions were performed for five days per week. Animal utilization and care were in accordance with the regulations and rules of the Ethical Committee for Animal Care and Utilization of the Science Research Center of Experimental Neurology, Republic of Georgia.

A Wisconsin general testing apparatus was used after modification for cats such that they can use the forelimbs to remove objects and obtain food. This apparatus, which has been described in detail in Okujava, Natishvili, Mishkin, Gurashvili, Chpashvili, Bagashvili, Andronikashvili, Kvernadze [2005], consists of two main sections: start cage for cats (60 x 65 x 60 cm) located on the experimental bench in a darkened, sound-proofed room, and a tray bearing three identical feeders, each of which was a round pot (depth 25 mm, diameter 73 mm) on which different objects (stimuli) could be placed. Stimuli consisted of a set of 600 toys of different sizes, shapes, textures and colors.

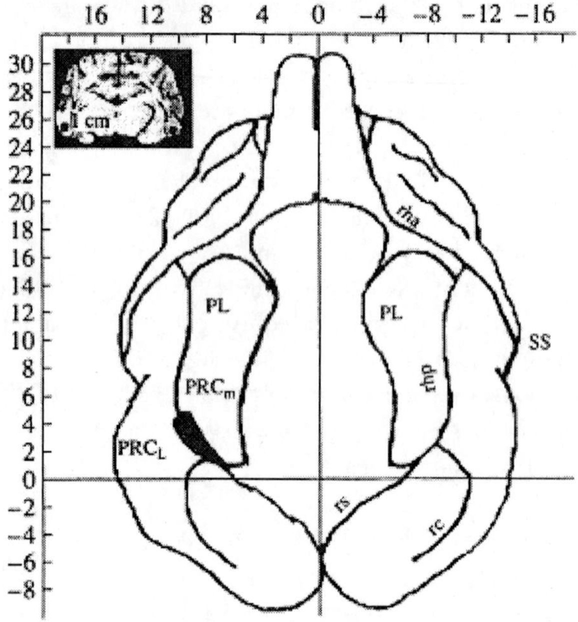

Figure 7. Diagram of the ventral surface of the cat brain. The black and grey areas show the target areas for lesions to the posterior parahippocampal area ["posterior rhinal cortex", as per Woznicka, A; Kosmal, 2003]. Numbers identify the coordinates of brain structures of the cat brain [Jasper, Ajmone-Marsan, 1954] on the ventral surface. PL is the piriform lobe, SS is the sylvian sulcus; rha is the anterior rhinal sulcus; rhp is the posterior rhinal sulcus; PRC_L is the posterior rhinal cortex, lateral part (grey); PRC_m is the posterior rhinal cortex, medial part (black); rc is the recurrent sulcus; rs is the retrosplenial sulcus. The panel at upper left shows a typical lesion produced by electrocoagulation of tissue on passage of a direct current of 2-5 mA through a single electrode.

Behavioral testing of the animals was performed in strict accordance with procedures used in studies of the same questioning primates [Malkova, Mishkin, 2003; Parkinson, Murray, Mishkin, 1988], with occasional exceptions which are identified in the text. The cats were initially trained to use the paw to remove the object covering one of the three feeders. These objects were not used again. Removal of an object from a feeder allowed the animal to see the food hidden within it (pieces of sausage of about 1 cm^3). After the cats had learned to remove the objects covering the feeders with ease, they progressed to formal testing, which consisted of three stages (stages 1–3) preceding a final stage (stage 4). At each stage, each trial consisted of two sequential elements – presentation of the object and its selection. The "delayed sample selection" version of the task was used at stage 1, in which both the position of the feeder and the object covering it served as additional memorization signals. The animal had to use the paw to remove a new object (sample) covering one of the three identical feeders not containing food. The position of the sample above one of the three feeders was randomized for sequential trials. After a 10-sec delay, the animal was presented with the object just presented at the selection phase, in the same position, but now accompanied by another, new object, covering one of the two remaining feeders in random order. The cat received reinforcement in the selection phase when it removed the object seen previously from the feeder in the same position perceived during the sample presentation phase in this trial. At stage 2, the animal perceived a new object covering an unreinforced feeder in the sample presentation phase, as in stage 1, but in this case was presented with the same object in the selection phase – the sample along with its identical copy covering one of the remaining three two feeders in random order. Thus, at this stage only the position of the sample could serve as a relevant stimulus to remind the animal, and it was rewarded when the sample was removed from the appropriate feeder. At each of these two initial stages, the cat received 20 trials on the experimental day, separated by intervals of 15 sec; there were five days of experiments per week and the procedure continued until the learning criterion was reached - a level of 90% correct responses on two sequential days of experiments.

Stage 3 consisted of three subtests, in each of which the sample presentation phase used different combinations of two of the three feeders. We started with the first subtest, in which the presentation phase used feeders 1 and 3, covered by different new objects (subtest "1-3", see Table 7). In this presentation phase of the test, the animal had to use the paw to remove both objects covering the empty feeders 1 and 3.

After a 10-sec delay, the animal was presented with one of the displayed objects and its general duplicate in the selection phase of the test, now covering either the same two feeders (i. e., feeders 1 and 3, type 1 tests) or one of the feeders covered in the first phase of the test (either 1 or 3) and the new feeder 2, which was not involved in the presentation phase (type 2 tests). Type 1 tests were tests for the "object-place association, while type 2 tests were simple "place" tests. Thus, in half the tests of this subset, tests were of the "object-place" type, in which the animal could perform correctly only by linking (associating) the object with its feeder place learned in the presentation phase. The other half of the tests were "place" tests, which the animal could solve correctly by remembering either the positions of the two object samples presented or the positions of the feeder not covered by an object in the presentation phase, avoiding this feeder in the selection phase. This procedure gave four different configurations of the tests for each of the three subtests. For clarity, these configurations are illustrated schematically in Table 7. Twelve tests were presented on the experimental day; testing was continued until the animal reached the 90% correct performance criterion in each of the three subtests on each of two sequential experimental days.

Table 7. Diagram representing twelve possible trial configurations in Stage 3 of formal testing. Symbols "A" and "B" represent different stimulus-objects covering respective food wells; on each successive trials different novel objects were used. Symbol "#" represents food wells used. "Plus" sign at the right of object represents rewarded response to it

Phases	Subtest 1 (feeders 1 -3)			Subtest 2 (feeders 1 -2)			Subtest 3 (feeders 2-3)		
	A		B	A	B			A	B
Sample presentation	#(1)	#(2)	#(3)	#*(1)	#(2)	#(3)	#(0)	#(2)	#(3)
phase									
	1)A(+)		A	1)A(+)	A		1)	A(+)	A
Selection phase	#(1)	(2)	#(3)	#(1)	#(2)	#(3)	#(1)	#2)	#(3)
	2)B		B(+)	2)B	B(+)		2)	B	B(+)
	#(1)	#(2)	#(3)	#(1)	#(2)	#(3)	#(1)	#(2)	#(3)
Tests for "place"									
	A		B	A	B			A	B
Sample presentation	#(1)	#(2)	#(3)	#(1)	#(2)	#(3)	#(1)	#(2)	#(3)
Phase									
	1)A(+)	A		1)A(+)		A	1) A	A(+)	
Selection phase	#(1)	#(2)	#(3)	#(1)	#(2)	#(3)	#(1)	#(2)	#(3)
	2)	B	B(+)	2)	B(+)	B	2)B		B(+)
	#(1)	#(2)	#(3)	#(1)	#(2)	#(3)	#(1)	#(2)	#(3)

Notes. The symbols "A" and "B" designate different stimuli – objects covering the corresponding feeders (1,2,3); new, different objects were used in each sequential trial. The symbol "#" identifies the feeders used ,a "+" to the right of the object shows reinforced reactions to it at the selection phase.

Finally, in the last (fourth) stage, tests consisting of all three third-step subtests were mixed with each other and presented in a random but balanced sequence. In each test session, two different sequences 0f 12 third-step trials were combined with each other such that half of the tests were of the "object-place" type and the other half were of the "place" type. The testing procedure was identical to that at the third stage. Testing was continued for 10 days (five days per week for two weeks); performance assessments averaged over the two weeks was used to evaluate the baseline level of pre-operative performance for each animal. For clarity, these procedures are illustrated schematically in Table 7.

After surgery, animals were tested only in the final (fourth) stage of the test for 30 days (five days per week for six weeks).

At the end of the preliminary training, the animals were randomized to two groups: a control group – all cats underwent surgical manipulations but without passage of the current through brain tissue. Cats of the second group underwent bilateral electrolytic lesioning of brain tissue in the area functionally equivalent to the posterior parahippocampal area in primates (the "postrhinal cortex" as defined by Woznicka & Kosmal (2003)). All operations were performed under deep anesthesia with Nembutal (40-60 mg/kg) in strictly aseptic conditions. Electrolytic lesions in this brain structure were produced by passing anode currents of 2-5 mA via stainless steel electrodes (tip diameter < 0.2 mm) insulated throughout with epoxy lacquer with the exception of a 0.5 mm length at the tip. The stereotaxic coordinates of brain areas subjected to electrolytic lesioning were determined using a stereotaxic atlas of the cat brain (Reinoso-Suarez 1961). Coagulation of the entire posterior parahippocampal area required insertion of the coagulating electrode at 2-3 different points into the brain through small openings (diameter 0.5-1 mm) drilled into the scull bone with a drill. In the trial experiments histological verification confirmed that the coagulation current parameters used here produced lesions of mean diameter 3-5 mm (see panel in Figure 7). After surgery was complete, wounds were closed, animals were given intravenous anti-edema agents and long-acting antibiotics and warmed with a heater. Post-operative testing started after 7-10 days.

After experiments were complete, cats were overdosed with i.p. Nembutal. Brains were perfused transcardially with normal physiological saline containing aldehyde fixatives, removed, photographed, and serial coronal sections of thickness 50 micrometer were cut on a freezing microtome; each fifth section was stained with thionine and covered with cover slips. Sections were monitored under the microscope to identify cell loss and gliosis and lesion sizes, which were plotted on standard diagrams of coronal sections. The sizes were measured in mm^2 for each section through the lesioned structure, measuring the margins of lesions using a simple digital flatbed system connected to a computer, with subsequent manual determination of lesion area.

Results and Discussion

Morphological monitoring of lesions

Brain tissue lesion sizes were expressed as percentages of the total volume of the structure of interest and results for each animal are shown in Table 8.

The data presented in Table 8 show that the posterior parahippocampal area was subjected to a significant level of destruction (mean 83%, range 73-93%) in all operated animals, while the hippocampus itself was not damaged, though significant bilateral involvement was seen in the area of the parasubiculum/presubiculum (mean 49%, range 33-56%, Table 8, Figure 8). The surrounding structures in all cats with parahippocampal damage also showed unintentional lesioning of the entorhinal and perirhinal cortical areas, though the level of this damage was no greater than 40%.

Table 8. Extent of Damage to Structures of Medial Temporal Lobe – Volume (mm^3), Side

Cat	HF (406)			Para/Presub (70)			ERh (213)			PRh (233)			Ph (150)		
	L	R	Mn	L	R	Mn	L	R	Mn	L	R	Mn	L	R	Mn
	0	0	0	61	51	56	28	23	26	26	20	23	75	70	73
	0	0	0	43	57	50	18	25	22	15	30	23	80	75	78
	0	0	0	32	34	33	15	20	18	10	15	13	87	85	86
	0	0	0	51	57	54	10	20	15	30	10	20	90	96	93
	0	0	0	47	50	49	18	22	20	20	19	20	18	17	83

Notes: HF is the hippocampal formation; Para/Presub is the parasubiculum + presubiculum; ERh is the entorhinal cortex; PRh is the perirhinal cortex; Ph is the parahippocampal area as defined in Woznicka & Kosmal (2003); L = left; R = right; Mn = mean.

Consequences of local lesions to the parahippocampal area.

These data are presented in Table 9.

There were no significant differences in performance of the “object-place” or “place” tests throughout pre-operative or post-operative testing in the control sham-operated or experimental animals with lesions to the parahippocampal area (pre-operative, Mann-Whitney U test: $m = n = 4$, $U = 7$, $p = 0.773$, two-tailed; post-operative testing, Mann-Whitney U test: $m = n = 4$, $U = 5$, $p = 0.468$, two-tailed). At the same time, a significant deficit in spatial recognition was seen after surgery between the experimental group of cats with parahippocampal lesions and the control group of sham-operated cats on performance of tests for spatial recognition (Mann-Whitney U test: m = n = 4, U = 0, p = 0.02, two tailed). This deficit was apparent to the same extent in the "object-place" and "place" tests.

Table 9. Pre-operative and post-operative training to perform the "place" and "object-place" tasks in cats with parahippocampal lesions (Ph1-Ph4) and sham-operated cats (SO1-SO$). Eight cats took part in pre-operative training. After they achieved the criterion of performing tests correctly , four cats underwent lesioning of the parahippocampal area and the others served as "sham-operated" controls. For the eight intact cats, the number of trials (T) and errors (E) for achievement of the 90% correct test performances criterion are shown

Cat	Preoperative training		Performance sessions 1-12			Post-operative performance sessions 1-12		
	T	E	OP	P	Mean	OP	P	Mean
Ph1	540	160	72	48	60	120	92	106
Ph2	500	125	60	68	64	98	102	100
Ph3	1220	244	96	100	98	100	82	91
Ph4	460	120	88	84	86	96	100	98
Mean	680	136	79	75	74	104	94	93
SO1	540	135	64	36	40	72	84	67
SO2	620	124	90	104	97	80	44	54
SO3	480	120	90	100	95	94	100	97
SO4	720	240	84	38	53	78	42	58
	590	155	100	70	71	81	68	112

Notes. "Ph" shows bilateral lesions to the parahippocampal area. SO = sham-operated; T = trials; E = errors to achievement of the criterion for pre-operative training including stages 1, 2, and 3. Each column shows the performance, averaged for two test sessions (12 configurations of trials, each including 144 trials) for the "object-place" (OP) and "place" (P) tests and means for the two types of test.

The data obtained here suggest: 1) that in cats, memory for the recognition of a one-trial perception of associations of the "object-place" type depend on the posterior parahippocampal area; 2) that within this area, both the parahippocampal cortex and the parasubiculum/presubiculum make significant contributions to organizing this type of memory.

Mishkin, Suzuki, Gadian, Vargha-Khadem [1997] suggested that memory for the recognition of different sensory modalities (including visual) is organized hierarchically. The hippocampal system receives its main inputs from widely distributed areas of the neocortical mantle via two main pathways: one running from the perirhinal area and other from the posterior parahippocampal area. The ventral information processing stream, important for organizing visual nonspatial recognition memory, includes the "higher visual areas" (the inferior temporal cortex), the perirhinal area, and the entorhinal cortex at the input to hippocampus itself. On the other hand, the same hypothesis identifies a dorsal information processing stream, providing for the organization of spatial recognition memory. In primates, this includes the inferior parietal cortical lobe and the posterior parahippocampal area at the input to hippocampus. Unfortunately, this question has not as yet been investigated in cats in this context. Thus, there is interest in recent neuromorphological data from cats which indicate that the rostrocaudal axis in the perirhinal cortex corresponds to the transition from the auditory to the visual areas in the temporal associative cortex [Witter, Groenewegen, 2004].

It is interesting to compare our data with Beritashvili's "image memory" concept in higher vertebrates, which is a natural extension of his views on the regulation of behavior in higher vertebrates by images of environmental objects occurring after single-trial perception of these objects [Beritashvili, 1971]. It should be noted that even at the very first phases of development, Beritashvili's concept arose from the following basic considerations: single-trial perception of the location of food in animals is followed by the creation of an image of the food location in the specific situation which can be reproduced after a period of time on repeated perception in the same situation – i.e., both the image of the food itself and the image of its location can be reproduced; it seems natural that the "food-place" association should also be reproduced. The analogy with the views of American investigators (at least in terms of the behavioral-functional aspect of this question) is almost complete. The only differences, from our point of view, relate to the following two points. Firstly, Beritashvili always emphasized the remembering of vitally important objects (such as food) and their locations in a defined situation (which is also remembered but, so to speak, "implicitly"). Secondly, and this is very important, in Beritashvili's studies, as in those of other authors at that time, the experimental data were not subjected to statistical analysis. At the same time, it should be noted that according to Beritashvili's views, only the neocortex is responsible for this type of image memory, while data published by American authors and our own results indicate that this type of cognitive memory is mediated by structures which are less part of neocortex than of the "intermediate cortex", as indicated by Filimonov (1949).

To summarize obtained results we can make following main inferences: 1.Intact cats can learn tests demonstrating recognition memory both for single-trial perception of associations between external objects and their spatial locations and for the single-trial perception of only the position of the object. 2. In cats, recognition memory for single-trial perception of associations of the "object-place" type depends on the posterior parahippocampal area; some contribution to this recognition is also made by the parasubiculum/ presubiculum complex, though the concrete role of the latter has yet to be established. 3. Recognition memory for single-trial perception of only the spatial position of an object also depends on the posterior parahippocampal area, with involvement of the parasubiculum/presubiculum complex.

And now to the end of our chapter we want to consider briefly from our point of view the delayed response problem – the problem, from which the original concept of I. Beritashvili about "image-driven" behavior has emerged [Beritoff, 1961].

IV. Two Subsystems of Spatial Visual Memory (Delayed Response Problem)

Now, let us return to more synthetic aspects of cognitive memory in animals. By "synthetic aspects" we mean here those aspects of memory research, which are closer to its manifestations in "real" life in "real" ecological conditions. Thus we need here to consider more ethological behavior than in artificial laboratory conditions, permitting stricter and more "analytical" research. We suppose in that respect that classical delayed response task introduced about a century ago in comparative psychology by Hunter [1913] might be appropriate for that purpose because at one hand it "models" natural behavior (for instance, food seeking behavior in animals – remember dog, which purposefully seeks the bone, hidden by it some times ago under earth), and on the other hand, in laboratory conditions permits quantitative, parametric analysis of behavior. As we have mentioned repeatedly it is our supposition, that Beritashvili in his memory research widely used such tasks like delayed response for its just mentioned advantages [Okujava, Natishvili 1986]. General logical structure of the delayed response (in its so called "direct version", (see [Fletcher, 1965]) is as follows:

An experimental animal is restricted in the start-cage with a transparent front door. An experimenter shows an animal the small part of food and at the full site of an animal hidden it in one of two (or several) identical foodwells. Just at the same time delay interval starts, after cessation of which an animal is released from the start-cage and its behavior is recorded; if it directly runs to the appropriate foodwell it is allowed to took the reward from it and to return back into the start-cage. This is description of one trial of the test. As a rule few trials per experimental day is given to an animal. Such trials are given to an animal in sufficiently large experimental situation, like those used by Konorski (Konorski, 1970) and Beritashvili (Beritashvili 1971). Figure 9 schematically represents test-situation used by Beritashvili for the study of the delayed responses in dogs and cats.

It is important to mention from this schematically view of situation, that it is quite inhomogeneous and reach of external landmarks, which might be used by the animal to localize the goal (food and its spatial location in situation).

How an animal placed in start-cage could spatially localize the food in such situation? Answer is evident – by the use of external landmarks. But this answer has an interesting alternative: what animal would do if situation became quite homogeneous, symmetrical and lacking any landmarks.

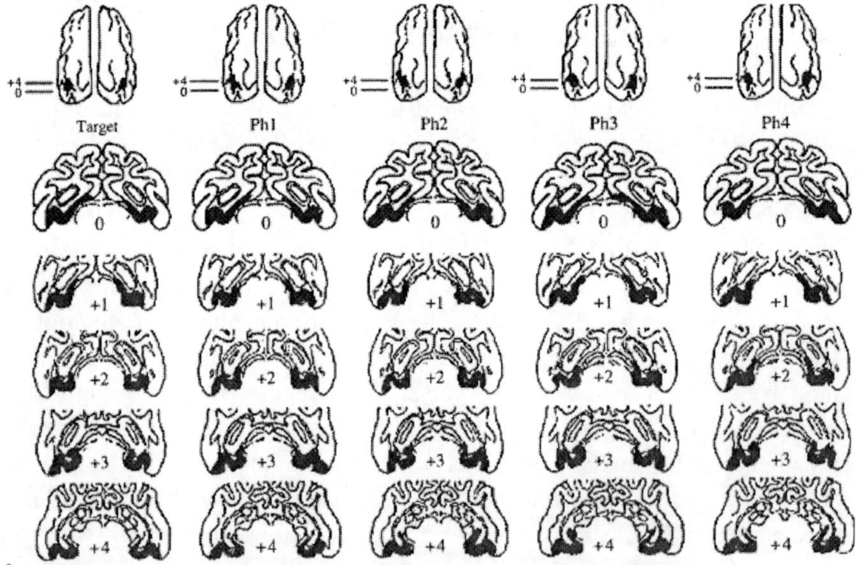

Figure 8. Target lesions of the posterior parahippocampal area (left column, designated "target") and actual lesions in four cats (Ph1-Ph4) are shown on standard coronal sections, which also show reconstructions on the standard ventral brain surface (upper row). Numbers identify the distance (mm) from the interaural vertical plane.

As is known from the theory of spatial function there exist two major systems for spatial localization – the egocentric and the allocentric systems (Figure 10).

In the first the spatial location of the stimulus of "interest" is defined in relation to the subjects own location (thus the term "egocentric"), while in the second spatial location of needed stimulus might be defined in relation to external landmarks (thus the term "allocentric"). (Bremner 1982). It is quite evident, that in first system spatial location of the "goal" is completely dependent on the subjects own spatial location (condition, for which the term "egocentic" is quite appropriate, as is the term "personal" system of spatial localization, which is preferred by us in future discussion); on the other hand, in the second system spatial location of the "goal" is in somewhat restricted conditions (see below) independent of subjects own movements (condition for which the term "allocentric" is quite appropriate, as is the term "extrapersonal" spatial localization, which is preferred by us in future discussion).

Now let us consider the delayed response in the context of these two systems of spatial localization. All the following experiments were performed on cats in somewhat artificial test-conditions, modeling in our view the above mentioned systems.

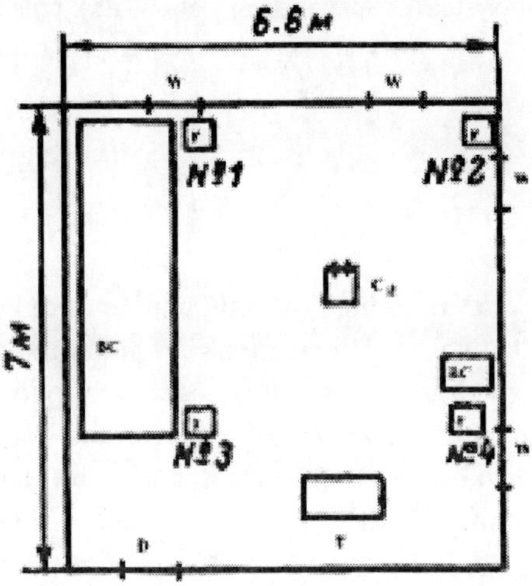

Figure 9. Situation in which all behavioral research of I. Beritashvili proceeded ; memory too was studied in that situation by the delayed response method. Note significantly allocentric (extrapersonal) character of such test-situation. Abbreviations: BC – bookcase, Cg – animal cage, D – door, F – food well, t – experimenter table, w – window

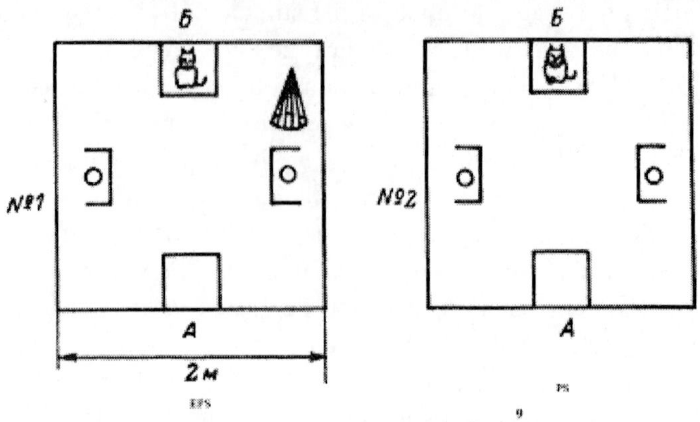

Figure 10. Situations in which our experiments with inverted delayed response proceeded. On the left test-situation to study delayed response in extrapersonal (allocentric) system of stimulus spatial localization (EPS); note the existence of external landmark (pyramid); on the right test-situation to study delayed response in personal (egocentric) system of spatial organization.(PS); note that in that case situation appears to be extremely homogenous and symmetrical, lacking any landmarks.

Materials and Methods

Experiments were performed on eight adult cats of both genders, weighing 2,8-3,5 kg. Experimental animals were placed in individual cages of size 1.5 x 1.0 x 1.0 m with free access to water. Food was provided daily at 20:00 prior to testing. Experimental sessions were performed for five days per week. Animal utilization and care were in accordance with the regulations and rules of the Ethical Committee for Animal Care and Utilization of the Science Research Center of Experimental Neurology, Republic of Georgia. Experimental situations in which animal testing proceeded were schematically depicted in Figure 10.

As can be seen from the right panel of this Figure, situation appeared quite homogenous and symmetrical, while in the left panel situation is made asymmetrical by introducing in it of one external object as a landmark. Thus we suppose that testing of the delayed response in the situation represented by the right panel will mean its testing in the "personal system" (PS) of spatial localization, while the same testing in the situation represented by the left panel will mean its testing in the condition of "extrapersonal system" (EPS) of spatial localization. Animal located in its start-cage was shown a small piece of boiled meat (approximately 1 cm^3, or the boiled sausage of the same size) and in the full site of the animal was hidden by the experimenter in one of two identical, symmetrically placed foodwells according to sequences given in Gellerman's table. Twenty trials of the above described type were given per day to animal; testing of the delayed response proceeded until an animal reached criterion of correct responses (at least 18 correct responses in 20 trial daily block). Four cats were firstly tested in delayed response in EPS situation, while after attaining criterion they were tested in Delayed response in the PS situation. Second group of four cats were initially tested in the delayed response in personal system situation and afterwards in the EPS situation. There was no significant difference between these two rules of testing (in terms of both trials and errors to criterion there was not significant difference between the two sequences of testing delayed response: first in EPS system and afterwards in PS system and *vice versa*). But test-situations depicted in Figure 10 are, of course, artificial; in real life animals continuously move and thus some kind of monitoring mechanism is needed for both systems of spatial localization.

In order to study this intriguing question in more detail we introduced the concept of the "inverted delayed response" (Natishvili 1978). In this modification of the classical delayed response an animal, sitting in the start-cage visually watches the food, which is hidden by an experimenter in one of two identical and symmetrically placed foodwells, after which delay interval starts. During this interval an animal in the cage was passively transported to a new place, symmetrically located relative to foodwells. After cessation of delay an animal is released from this new place of the start-cage. As is evident from pure imagination, at the moment of release from the start-cage test situation before animals view is a mirror image of situation which existed at the moment of its perception.

Thus, if intact cats were able to solve correctly this "inverted delayed response" task, it will mean that they possess some brain mechanism capable of continuous monitoring of the "inversion procedure" (that is passive transportation during the delay to a new starting point, from which they will be released after cessation of the delay). What kind of sensory information might be used for such monitoring? No doubt, visual and vestibular information might be the only candidates. We have excluded visual information by making the start-cage completely opaque; thus the only source for correction might be the vestibular information. This simple supposition was experimentally documented by us in the following simple experiment, the purpose of which was to induce during passive transportation the highly noisy information in the vestibular feedback system.

Experiments were performed on the same cats used in just preceding experiment with all procedures of testing identical with the exception of the "inversion operation" which consisted in rapid clockwise and anticlockwise rotations of a passively transported animal in opaque cage during some moment of delay interval.. Theoretically one could predict that in PS situation cats will fail to solve correctly this task, while in EPS situation their solving of the task will be deranged, but without complete fail.

Data obtained are shown in the Table 10. One could see with ease from these data that cats in inverted delayed response being tested in the PS situation demonstrated erroneous responses when rotated during inversion operation; it was important that errors in that case were quite predictable in many if not all cases. When vestibular feedback was destroyed by the noise, but memory still existed in experimental subjects an animal's response uncorrected by that feedback oriented it to the foodwells perceived before inversion operation during delay period.

But what brain structure mediates the delayed response performance in normal animals – representatives of many different species?

Approximately from the pioneering studies of Jacobsen in 30-ties of past century it is well known that quite selective and long-lasting deficit in delayed response arises after prefrontal lesions [Jacobsen, 1935]. This research has induced a lot of neuropsychological studies performed mainly in primates, rodents and carnivores which not only confirmed initial observation of Jacobsen, but led to further delimitations of the critical areas involved in the production of the "frontal lobe syndrome". Because our own research is limited to cats as representatives of carnivores we shall briefly review here the prefrontal cortex of cats and the situation with delayed response.

The definition of the prefrontal cortex as the projection zone of the mediodorsal nucleus of the thalamus is based on the work of Rose and Woolsey [1948] who showed that this nucleus projects to anterior and ventral parts of the brain in nonprimates. Rose and Woolsey however termed this projection zone "orbitofrontal." It seems to have been Akert [1964] for the first time explicitly suggested that this criterion could be used to define homologues of the prefrontal cortex in primates and nonprimates. This allowed the establishment of homologies despite the lack of a granular frontal cortex in nonprimates.

Table 10. Errors in performance of the inverted delayed response (IDR) are shown in normal cats after achievinged criteria level of performance in this task in two situations of spatial localization of stimulus to be remembered – personal system (PS) and extrapersonal system (EPS) under two conditions of testing: normal and with vestibular interference. Errors made in 10 consecutive trials of the IDR are indicated.

Cats	Errors in IDR in PS situation	
	Performance at criteria level	Performance after vestibular interference
#1	0	4
#2	2	5
#3	1	6
#4	0	7
Cats	Errors in IDR in EPS situation	
	Performance at criteria level	Performance after vestibular interference
#1	1	3
#2	0	1
#3	0	2
#4	2	3

In primates prefrontal cortex is frequently viewed as a "granular frontal cortex" because it contains a lot of small granular cells in one of its cortical layer [Nauta, 1972]. Such layer is completely absent or undeveloped in carnivores and rodents; nevertheless "prefrontal cortex" in those species might be defined if we instead of sufficiently unreliable cytoarchitectonic criteria used other neuromorphological criteria for the percolation of the neocortex in these sub primate species as have been done by Rose, Woolsey, Akert and their followers.

As to behavioral functions of the prefrontal cortex in primates M. Mishkin for delayed alternation and N. Butters for delayed response have shown that with correct performance of the classical delayed response and delayed alternation is precisely connected only small part of the prefrontal cortex – namely s. principalis and within it especially its middle third [Mishkin, 1957; Butters, Pandya, Stein, Rosen, 1972]. On the other hand behavioral studies of the delayed response in carnivores were performed by I. Beritashvili and J. Konorski [Beritashvili, 1971; Konorski, 1967]. Those studies have shown that both in dogs and cats substantial deficits in delayed response performance results following bilateral lesions of gyrus proreus – region in dorsolateral prefrontal cortex in carnivores, functionally analogous to sulcus principalis in primates Having in mind these brief considerations we decided to examine the possible functional role of proreal gyrus in both classical and so called inverted delayed response.

Results and Discussion

Nine cats were tested in the inverted delayed response in PS system of stimulus spatial localization; among these nine cats six had bilateral parieto-temporal lesions (subpial aspirations) in the middle suprasylvian gyrus, while three cats had bilateral subpial aspirations in the gyrus proreus (Figure 11).

On the other hand three cats with bilateral proreal lesions and four cats with bilateral lesions of the parieto-temporal lesions were tested in the inverted delayed response in the EPS situation of spatial localization. Twenty trials per experimental day were given to our cats in learning classical and inverted delayed response under two conditions of testing – in PS and EPS situations. Learning in lesioned and normal cats proceeded until their performance reached the criteria level – no more than 2 errors during 20 consecutive trials (one experimental day). Testing situations for the two systems of spatial localization (PS and EPS) was already described above.

Results of these two experiments (performance of inverted delayed response in normal and operated cats in two testing situations are shown in Figure 12).

As evident from the data presented in this Figure, cats with prefrontal damage (bilateral lesions of g. proreus) exhibited significant deficits in inverted delayed response in PS system of spatial localization (m=3, n=6, U=0, P=0,01, Mann-Whitney). On the other hand cats with parietal lesions (bilateral lesions of posterior suprasylvian gyrus) exhibited significant deficits in inverted delayed response performance in EPS system of spatial localization. In this context it seems interesting to us to compare our data with the results of American neuropsychologist Pohl [1973], results obtained in monkeys with prefrontal and parietal lesions. He has studied in monkeys quite a different behavioral test – spatial reversals in two testing situations – one symmetrical and homogenous (PS situation) and the other situation with external visual landmarks (EPS situation). Pohl found double dissociation of behavioral deficits in monkeys after prefrontal and parietal lesions: prefrontal monkeys exhibited deficits in spatial reversals in the behavioral situation lacking any landmarks (e.g. PS system), while parietal monkeys were deficient in spatial reversals in behavioral situation with landmarks. These observations are very important because of the following reasons: a) they accentuate the possible linkage of the dorsolateral prefrontal cortex to the "egocentric spatial functions" (although the spatial reversal learning and the classical delayed response task are different behavioral tasks, their linkage to the dorsolateral prefrontal cortex in mammals is highly well documented [e. g. Rosvold, 1972]; b) they point to parietal cortex in monkeys to be responsible for "allocentric spatial function"(although one must be very precautious in postulating functional "analogy" between well-studied parietal cortex in primates and sufficiently less well-studied functions of parietal cortex in carnivores (cats in our case).

Now we want to widen our discussion in such a way as to encompass some more general views concerning functions of the prefrontal cortex in mammals. First let us begin with the well-known hypothesis of so called "corollary discharge" developed by Hans-Lukas Teuber [Teuber, 1972]. Briefly speaking according to this hypothesis any self-produced, active volitional movement in the animal is accompanied by such "intracerebral" discharge, which prepares the sensory-perceptual brain systems to those changes in afferent stimulation which will be the result of such movements. Of course in our version of the inverted delayed response start-cage with an animal was passively transported to a new, symmetrically situated place and thus "corollary discharge" could not participate in compensation process needed to "mentally update" initially perceived food well after completion of inversion procedure; in that case vestibular feedback, to our mind, wood be sufficient for such updating. But in "real life", in natural conditions animals move actively and in such cases vestibular feedback functions in parallel with "corollary discharge". In general this brief consideration points to possible links between psychological personality and biology of prefrontal cortex. Interestingly enough from the neuroanatomical and neuropsychological points of view prefrontal cortex appears to be "association cortex" for limbic system of the brain [Nauta 1964, Pribram, 1967] – in other words, the processes mediated by prefrontal cortex of the brain are related to "primitive", simple visceral sensations in such a way as complex perceptual representations (mediated by sensory "association" areas) are related to simple, "primitive" sensations, mediated by primary sensory areas of the brain.

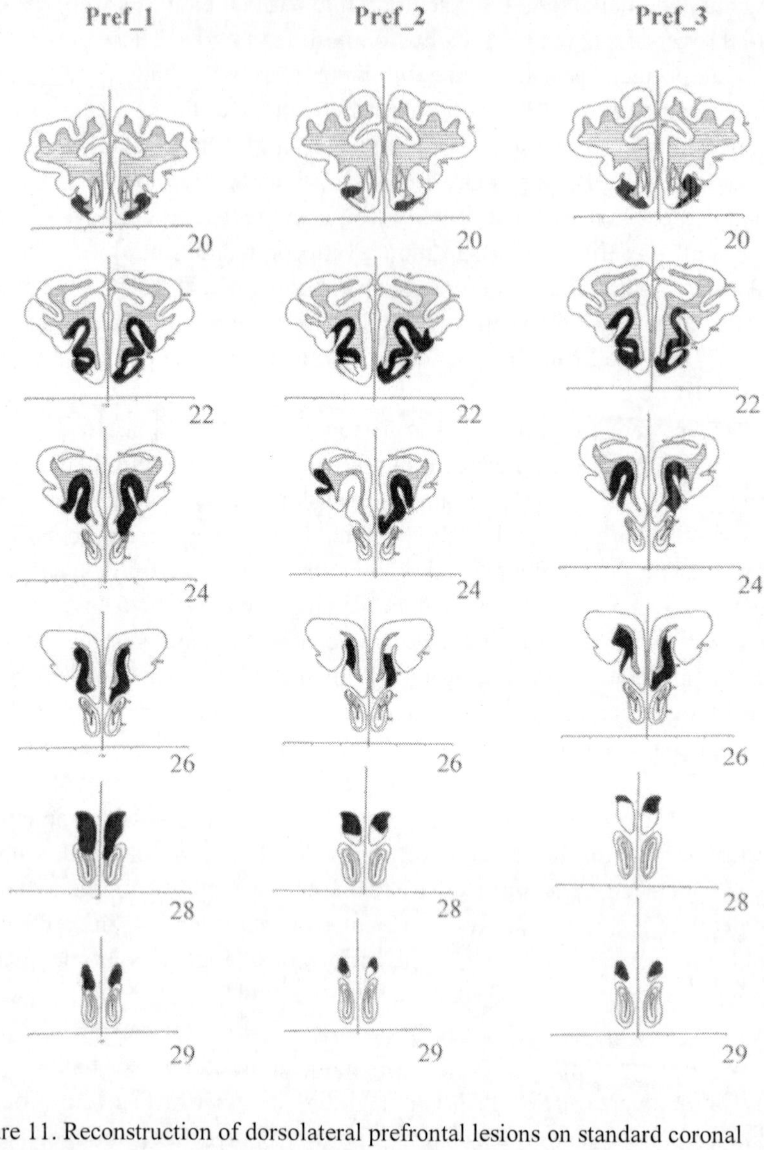

Figure 11. Reconstruction of dorsolateral prefrontal lesions on standard coronal sections in three cats participated in inverted delayed response testing, represented in Figure 12. Numbers identify the distance (mm) from vertical interaural plane. Sections are from the stereotaxic atlas of the prefrontal cortex of the cat after H. J. Markowitsch and M. Pritzel [1977].

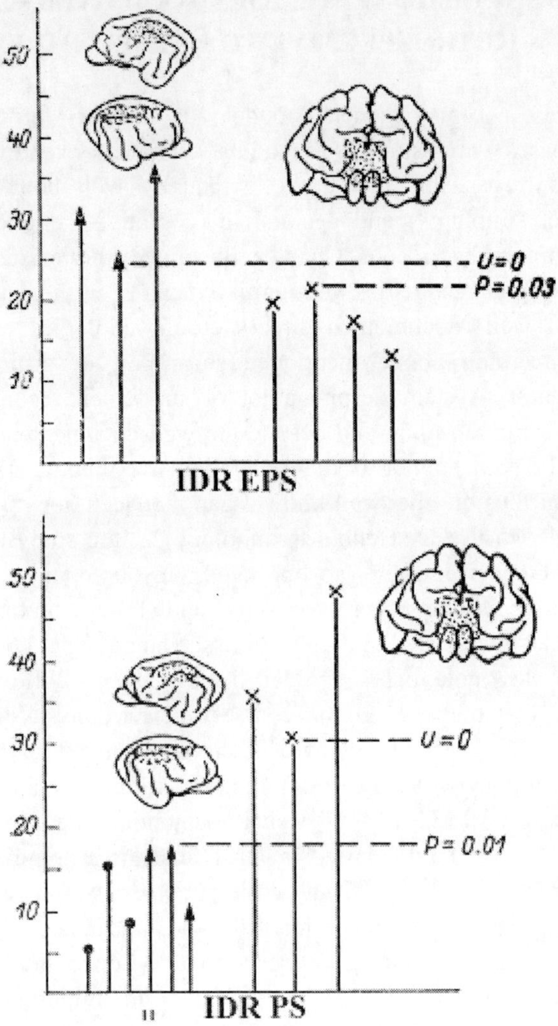

Figure 12. Double dissociation of deficits in inverted delayed response (IDR) in cats after selective lesions of prefrontal and parietal association cortices in the two testing situations (EPS - extrapersonal system vs. PS - personal system of spatial localization) following prefrontal (indicated by crests) vs. parietal lesions (indicated by dashed triangles); dashed circles indicate behavior of sham-operated cats (with ought lesions of the cortical tissue). Mann-Whitney U test for statistical comparisons. Lesion areas are indicated by dots on the cat's dorsolateral cortex.

V. CONCLUSION: MEMORY REVISITED – FROM RETROSPECTIVE TO PERSPECTIVE

Commonly accepted general model for memory processes includes following phases – registration (encoding of trace), retention and retrieval (decoding of trace); this model works equally well both for procedural (noncognitive, "implicit") and declarative (cognitive, "explicit") memories. Preceding paragraphs of our Chapter illustrated the existence and some neurobehavioral characteristics of cognitive memory in the cat. Unfortunately the deficit of well- documented direct neuropsychological evidence is yet observed concerning above mentioned separation of different phases in cognitive memory system despite a lot of single-cell recordings from the neurons in the rhinal region and despite the well-documented behavioral and psychological data in support of that separation. We think that major reason of such situation lies in the well-known gap between the "molecular" and "molar" levels (to use D. Hebb's terminology) in the study of memory. To jump successively over such gap one needs, in our opinion, to select such parameter, which will be "macroscopically" and at the same time will in some way reflect the "microstate" of the neural tissue being under consideration. We consider the "slow potentials" recorded from the rhinal cortex to be possible candidate for that purpose. In the course of electrophysiological study we performed recording of slow electric potentials from the cat's rhinal region during different phases of the delay (initial 5 s, middle 5 s and final 5s) in the both DMS and DNMS tasks of visual recognition memory under testing conditions with 15 second delays. Statistical analysis of the data obtained has shown that statistically significant correlation exists between the correct behavioral performance of recognition tasks and appearance of slow potentials in the rhinal region only during last seconds of the delay interval, but not for its initial and middle phases. This result points to the importance of the rhinal region for recall process in the recognition memory (Okujava et al., 2009). Additional although indirect data pointing to the involvement of the rhinal cortex in the cats in retrieval-reproduction process may be inferred from the Table 6 (see II. Nonspatial visual recognition in cats), in which single-factor ANOVA demonstrated a statistically significant intergroup difference between control and experimental cats with rhinal lesions to the brain ($F = 5.935$; $p = 0.009$). On the other hand, the same analysis showed that the post hoc Sheffe's test for multiple comparisons identified a significant difference between cats with lesions to the rhinal area and animals of the control group with a 10-sec delay ($p = 0.024$) but not with a 5-sec delay ($p = 0.999$).

We began this Chapter with brief consideration of the I. Beritashvili's work in the field of animal behavior with stressing his impact in the field of cognitive memory (see Introduction). Now, at the final section we want to ask our reader not to be very critical to our somewhat pretentious attempt to reconsider Beritashvili's "image-driven memory" from more modern position; at least we wanted to fill his theory with strict statistical ground, which he mistakenly discarded. His work reminds us famous Georgian painter Niko Pirosmani, who being an artist-primitivist nevertheless attracted great interest all over the world up to now.

References

Adey, WR. Intrinsic organization of cerebral tissue in alerting, orienting and discriminative responses. In: Quarton GC; Melnechuk P, Schmitt F O (Eds.). *Neurosciences-A study program*. New York: Rockefeller Univ. Press; 1967, 615-633.

Akert, K. Comparative anatomy of frontal cortex and thalamofrontal connections. In: JM; Warren, Akert (Eds). *The Frontal Granular Cortex and Behavior*. New York: McGraw-Hill; 1964, 372-397.

Anokhin, PK. *Biology and Neurophysiology of the Conditioned Reflex*. Moscow: Nauka; 1979 (in Russian).

Ashby, WR. *Design for a Brain*. London: Chapman & Hall Ltd., 1960.

Beritashvili, IS. *Vertebrate Memory, its Characteristics and Origin*. New York: Plenum Press; 1971.

Beritov, IS. Individually Acquired Activity of the Central Nervous System. Tiflis: *Gosizdat*; 1932 (in Russian).

Beritoff, IS. Neural Mechanisms of Higher Vertebrate Behavior. Boston: Little, *Brown and Company*, 1965.

Beritov, IS. Structure and Function of the *Cerebral Cortex. Moscow*: Nauka; 1969 (in Russian).

Bernstein, NA. Descriptions in the Physiology of Motion and *Physiology of the Activity*. Moscow: Medicina; 1966 (in Russian).

Bremner, JG. Egocentric versus allocentric spatial coding in nine-month-old infants: factors influencing the choice of code. *Dev. Psychol.*, 1978, 14, 346-355.

Brodmann, K. Vergleichende Localisationslehre der Grosshirnrinde in ihren *Prinzipien dargestellt auf Grund des Zellenbaues. Leipzig*. JA. Barth; 1909.

Burwell, RD. Borders and cytoarchitecture of the perirhinal and postrhinal cortices in the rat. *J. Comp. Neurol.*, 2001, 437, 17-41.

Butters, N; Pandya, D; Stein, D; Rosen, J. A search for the spatial engram within the frontal lobes of monkeys. *Acta Neurobiol. Exp.* 1972, 32, 305-329.

Callahan, H; Ikeda-Douglas, C; Head, E; Cotman, CW; Milgram, W. Development of a protocol for studying object recognition memory in the dog. Progr. *Neuropsychopharmacol. Biol. Psychiat.*, 2000, 24, 693-707.

Campbell, Jr. Deficits in visual learning produced by posterior temporal lesions in cats. *J. Comp. Physiol. Psychol.*, 1978, 92, 45-57.

Filimonov, KN. Comparative Anatomy of the Cerebral Cortex in Mammals: the Paleocortex, the Archicortex, and the *Diencephalic Cortex. Moscow*: Medgiz; 1949 (in Russian).

Fletcher, HJ. The delayed-response problem. In: Schrier AM, Harlow HF, Stolnitz F (Editors). *Behavior of Nonhuman Primates*. New York: Academic Press; 1965, 129-169.

Gaffan, D. Recognition impaired and association intact in the memory of monkeys after transaction of the fornix. *J. Comp. Physiol. Psychol.*, 1974, 86, 1100-1109.

Gerard, RW. The fixation of experience. In: A; Fessard, RW; Gerard, J; Konorski, (Eds). *Brain Mechanisms and Learning*. Oxford: Blackwell; 1961, 21-32.

Gross, CG. Visual functions of inferotemporal cortex. In: R; Jung, (Eds.), *Handbook of Sensory Physiology*: V. 7/3B: Central processing of visual information. Berlin: Springer-Verlag; 1973, 451-482.

Hara, K; Cornwell, PR; Warren, JM; Webster, IH. Posterior extramarginal cortex and visual learning by cats. *J. Comp. Physiol. Psychol.*, 1974, 87, 884-904.

Hodos, W; Campbell, CBG. Scala Naturae: Why there is no theory in comparative psychology. *Psychol. Rev.* 1969, 78, 337-350.

Hollander, M; Wolfe, DA. *Nonparametric Statistical Methods.*, New York: John Willey and Sons; 1973.

Hunter, WS. The delayed reaction in animals and children. *Animal Behav. Mono.*, 1913, 2, 1-86.

Jacobsen, CF. Studies of cerebral function in primates. I. The function of the frontal association areas in monkeys. *Comp. Psychol. Monogr.*, 1936, 13, 3-60.

Jasper, HH; Ajmone–Marsan, CA. Stereotaxic Atlas of the Diencephalon of the Cat. Ottawa: *Natl. Res.* Council of Canada. 1954.

Katz, B. *Nerve, Muscle and Synapse*. New York: McGraw-Hill; 1966.

Klatzki, RL. *Human Memory: Structure and Processes*. San Francisco: W. H. Freeman and Co. 1975.

Konorski, J. Integrative Activity of the Brain. Chicago and London: *The University of Chicago Press*, 1967.

Kreiner, J. The neocortex of the cat. *Acta Neurobiol.* Exp., 1971, 31, 151-203.

Krettek, JE; Price JL. Projections from the amygdaloid complex to the cerebral cortex and thalamus in the rat and cat. *J. Comp. Neurol.*, 1977, 172, 687-722.

Krushinskiy, LV. *Biological Basis of Rational Activity in Animals*. Moscow: Moscow University Press; 1966, (in Russian).

Malkova, L; Mishkin, M. One-trial memory for object-place associations after separate lesions of hippocampus and posterior parahippocampal region in the monkey. *J.Neurosci.*, 2003, 23, 1956-1965.

Meunier, M; Bachevalier; J; Mishkin, M; Murray EA. Effects on visual recognition of combined and separate ablations of the entorhinal and perirhinal cortex in rhesus monkeys. *J. Neurosci.*, 1977, 13, 5418-5432.

Markowitsch, HJ; Pritzel, M. A stereotaxic atlas of the prefrontal cortex of the cat. *Acta Neurobiol. Exp.*, 1977, 37, 63-81.

Mishkin, M; Delacour, J. An analysis of short-term visual memory in the monkey. J. Exp Psychol. *Anim. Behav*. Process., 1975, 1, 326-334.

Mishkin, M; Murray EA. Stimulus recognition. Curr. Opin. *Neurobiol.*, 1994, 4, 200-206.

Mishkin, M; Prockop, ES, Rosvold, HE. One-trial object-discrimination learning in monkeys with frontal lesions. *J. Comp. Physiol. Psychol.*, 1962, 55, 178-181.

Mishkin, M. A memory system in the monkey. Philos. Trans. R. Soc. *Lond. B Biol. Sci.*, 1982, 298, 85-95.

Mishkin, M. Effects of small frontal lesions on delayed alternation in monkey. *J. Neurophysiology.*, 1957, 20, 615-622.

Mishkin, M; Suzuki, WA, Gadian, DG; Vargha–Khadem F. Hierarchical organization of cognitive memory. Philos. *Trans. R. Soc. Lond. B. Biol.*, S. 1997, 352, 1461-1467.

Mumby, DG; Pinel, JPJ. Rhinal cortex lesions and object recognition in rats. *Behav. Neurosci.*, 1997, 108, 11-18.

Murray, EA. Medial temporal lobe structures contributing to recognition memory: The amygdaloid complex *versus* rhinal cortex. In: Aggleton JP (ed). *The Amygdala: Neurobiological Aspects of Emotion, Memory and Mental Dysfunction*. London: *Wiley-Liss*, 1992, 287-296.

Murray, EA; Bachevalier, J; Mishkin, M. Effects of rhinal cortical lesions on visual recognition memory in rhesus monkeys. *Soc. Neurosci. Abs*., 1989, 15, 342.

Murray, EA; Bussey, TI. Perceptual–mnemonic functions of the perirhinal cortex. *Trends Cogn. Sci*., 1999, 3, 142-151.

Natishvili, T. Concerning one modification of the delayed response. *Bulletin of the Georgian Academy of Sciences*., 1979, 93, 161-164 (in Russian).

Natishvili, TA. Some Results of Neuropsychological Study of Animal's Memory. Neurophysiological Bases of Memory. In: Oniani TN (ed), *Neirophysiological bases of memory*. Tbilisi: Metsniereba, 1979, 378-398.

Nauta, WJH. Some efferent connections of the prefrontal cortex in the monkey. In: *The Frontal Granular Cortex and Behavior*. New York: *McGraw-Hill*, 1964, 397-407.

Nissen, HW; Riesen, AH; Nowlis, V. Delayed response and discrimination learning by chimpanzees. *Journal of Comparative Psychology*, 1938. 26, 361-386.

Okudjava, VM; Natishvili, TA; Gurashvili, TT; Chipashvili, SA; Bagashvili, TI; Andronikashvili, GT; Kvernadze, GG; Gogeshvili, KSh; Okujava, MV. Slow potentials in the rhinal region of the cat brain cortex related to visual recognition memory. Neurophysiology, 2009, 4, 327-335.

Okujava, V; Natishvili, T. On the problem of the behavioral act. Zh. Vysh. *Nerv. Deiat*, 1986, 36, 1156-1167 (in Russian).

Okujava, V; Natishvili, T; Mishkin, M; Gurashvili, T; Chipashvili S; Bagashvili T; Andronikashvili G; Kvernadze G. One-trial visual recognition in cats. *Acta Neurobiol. Exp*., 2005, 65, 205-212.

Olton, DS. Spatial memory. *Sci. Amer*., 1977, 236, 82-98.

Parkinson, JK; Murray, EA; Mishkin, M. A selective mnemonic role for the hippocampus in monkeys: memory for the location of objects. *J. Neurosci*., 1988, 8, 4159-4167.

Passingham, RE Primate specialization in brain and intelligence. *Symp. Zool. Soc*. Lond., 1981, 46, 361-388.

Pohl, W. Dissociation of spatial discrimination deficits following frontal and parietal lesions in monkeys. *J. Comp. Physiol. Psychol*., 1973, 82, 227-239.

Pribram, K. The limbic systems, efferent control of neural inhibition and behavior. In: WR; Adey, T; Tokizane, (Eds.), *Progress in Brain Research*. Amsterdam: Elsevier Publishing Co.; 1967, 27, 318-336.

Reinoso-Suarez, F. Topographysches Hirnatles der Katze fur experimental-physiologische Untersuchungen. Herausgegeben von E. *Merck A.g. Darmstadt.*, 1961.

Rose, JE; Woolsey, CN. The orbitofrontal cortex and its connections with the mediodorsal nucleus in rabbit, sheep and cat. Res. Publ. Assoc. *Res. New. Ment. Dis.*, 1948, 27, 210-232.

Rosenblueth, A; Wiener, N, Bigelow J. Behavior, Purpose and Teleology. *Philosophy of Science*, *Baltimore*, 1943, 10, 18-24

Rosvold, HE; The frontal lobe system: cortical-subcortical interrelatioships. *Acta Neirobiol. Exp.*, 1972, 32, 439-460.

Squire, LR. Declarative and nondeclarative memory: multiple brain systems supporting learning and memory. *J. Cogn. Neurosci.*, 1992, 4, 232-243.

Squire, LR; Zola-Morgan, S. The medial temporal lobe memory system. *Science*, 1991, 253, 1380-1386.

Steckler, T; Drinkenburg, WHI; Sahgal, A; Aggleton, JP. Recognition memory in rats – I. Concepts and classification. *Prog. Neurobiol.*, 1998, 54, 289-311.

Teuber, HL. Unity and diversity of frontal lobe functions. *Acta Neurobiol. Exp.*, 1972, 32, 615-656.

Tolman, EC. Purposive Behavior in Animals and Men. New York: *Appleton-Century-Crofts*, 1932.

Von Neumann, J. The general and logical theory of automata. In: L. Jeffres (ed). *Cerebral Mechanisms in Behavior*. New York: John Wiley and Sons, Inc., 1951, 1-31.

Warren, JM; Warren, HB, Akert, K. The behavior of chronic cats with lesions in the frontal association cortex. *Acta Neurobiol. Exp.*; 1972, 32, 361-392.

Warren, JM. Primate learning in comparative perspective In: Schrier AM, Harlow HF, Stollnitz A (eds). *Behavior of Nonhuman Primates*. New York: Academic Press; 1965; 249-281.

Witter, MP; Groenewegen, HJ. Connections of the parahippocampal cortex in the cat. III. Cortical and thalamic efferents. *J. Comp. Neurol.*, 2004, 252, 1-31.

Witter, MP; Groenewegen, HJ; Lopes da Silva, FH, Lohman, AHM. Functional organization of the extrinsic and intrinsic circuitry of the parahippocampal region. *Prog. Neurobiol.*, 1989, 33, 161-253.

Woznicka, A; Kosmal, A. Cytoarchitecture of the canine perirhinal and postrhinal cortex. *Acta Neurobiol. Exp.*, 2003, 63, 197-209.

Zola-Morgan, S; Squire, LR; Rasmus, NL. Severity of memory impairment in monkeys as function of locus and extent of damage within the medial temporal lobe memory system. *Hippocampus*, 1994, 4, 483-495.

In: Spatial Memory
Editor: Jacob B. Thomas, pp. 51-75
ISBN: 978-1-61668-139-5

Chapter 2

SPATIAL MEMORY: THE ROLE OF EGOCENTRIC AND ALLOCENTRIC FRAMES OF REFERENCE

Gennaro Ruggiero*[*], *Tina Iachini, Francesco Ruotolo and Vincenzo Paolo Senese
Department of Psychology, Second University of Naples, Italy

ABSTRACT

A fundamental requirement of everyday life is that of encoding and remembering successfully the locations of objects, landmarks or buildings in space. This function is achieved by structuring spatial information in systems of coordinates. There are many ways to classify spatial reference systems (Paillard, 1991), but a useful one for understanding human spatial memory, divides them into two main categories. Egocentric frames of reference specify location and orientation with respect to the organism, and include eye, head, and body coordinates. Allocentric frames of reference specify location and orientation with respect to elements and features of the environment independently of the viewer's position.

Given the primary role that egocentric and allocentric processing systems play in perception and action (e.g. Milner & Goodale, 1995,

[*] Corresponding author: Department of Psychology, Second University of Naples, Via Vivaldi 43, 81100 Caserta, Italy, e-mail: gennaro.ruggiero@unina2.it

2008; but see Schenk, 2006), they can be thought of as specialized cognitive mechanisms essential in performing and guiding spatial motor actions. Egocentric representations have a special relevance in controlling movement in surrounding space such as avoiding obstacles or reaching objects. All these actions are performed in near/peripersonal space, i.e. the space within arm-reaching distance, and require fine-grained metric information. On the other hand, allocentric representations have an important role in recognizing objects, scenes and planning future movements (i.e. the space outside arm-reaching distance) (e.g. Milner & Goodale, 2008). Spatial memory is intrinsically linked to frames of reference as it is not possible to store spatial information without structuring it according to specific frames. In this chapter we review evidence from behavioral and neurofunctional data about models of spatial memory, automatic and effortful encoding of spatial information, and spatial memory in elderly people. Furthermore, two experiments are reported that investigate memories for egocentric and allocentric frames of reference by means of a task requiring judgments of relative distance. Experiment 1 investigates the influence of near/far spaces on spatial frames of reference. Experiment 2 controls the influence of possible artifactual effects. Overall, the pattern of data shows that the egocentric processing is more accurate and faster than the allocentric one. The results are discussed in relation to models of spatial memory that emphasize the importance of egocentric experience (Kosslyn, 1994; Millar, 1994).

INTRODUCTION

A fundamental requirement of our daily life is spatial ability, that is the ability of encoding and remembering successfully the locations of objects, landmarks or buildings in space. In the literature, the definition of "spatial ability" is to some extent ambiguous as it has been used with different meanings and has been considered in a variety of ways. For example, spatial ability is associated with the processing of geometric (or metric) properties such as distance and size, as well as dynamic properties such as velocity and strength. The ability to encode the characteristics of objects such as size, orientation and location is also defined "spatial" (Pinker, 1984). Finally, the ability to navigate in the environment is also considered "spatial" because it requires an understanding of all these properties (Eilan, 1993).

Potentially, all kinds of processes and information useful to locate positions and directions in the environment can be defined spatial. The location and orientation of an object cannot be encoded without establishing a frame of reference, that is a coordinate that acts like anchor-point. Human memory must also use frames of reference to specify the locations of objects. Indeed, spatial information is structured in systems of coordinates within which spatial positions are represented.

In the domain of spatial cognition, spatial reference systems used to encode and organize in memory spatial information are commonly divided into two main categories: egocentric and allocentric (Kosslyn, 1994; Levinson, 1996; O'Keefe & Nadel, 1978; Paillard, 1991; Pani & Dupree, 1994; Piaget & Inhelder, 1967; for a review see Avraamides & Kelly, 2008).

Egocentric frames of reference use the organism as the center of the organization of surrounding space: retinotopic coordinates, head-centered and body-centered frames of reference may serve like anchor-points (Diwadkar & McNamara, 1997; Franklyn & Tversky, 1990; Kosslyn, 1994; Roskos-Ewoldsen, McNamara, Shelton, & Carr, 1998; Shelton & McNamara, 1997). Being sensitive to the subjective vantage point, egocentric spatial representations are often defined as "orientation-specific" or "orientation-dependent" (e.g. Presson, DeLange, & Hazelrigg, 1989). Consequently, access to spatial locations is biased by the relation between the required location and the organism.

Allocentric frames of reference have been defined in various ways, although they share a common aspect: spatial information is specified independently of the organism's position (O'Keefe & Nadel, 1978). According to some theories, this implies that all spatial positions in the environment are equiavailable and derived spatial representations are called "orientation-free" or "orientation-independent" (Rieser, 1989; Waller, Montello, Richardson, & Hegarthy, 2002). Other theories highlight the importance of external objects chosen like anchor-points, such as: objects or parts of objects (Humphreys & Riddoch, 1994; Kosslyn, 1994), salient landmarks, local features like walls and global features like mountains (McNamara, 2003; McNamara, Rump, & Werner, 2003), intrinsic axes defined by inter-objects relations (Mou & McNamara, 2002), the Sun azimuth and the direction of gravity (Paillard, 1991). For these reasons, allocentric frames of reference are also called "environmental" or "geocentric" frames (see McNamara, 2003).

Some influential theoretical models assign a primary role to egocentric and allocentric processing systems in perception and action (e.g. Milner & Goodale, 1995, 2008; Schenk, 2006). Egocentric representations have a special relevance in controlling movement in surrounding space under perceptual guidance; for example, they are essential to avoid obstacles or to reach objects. All these actions are performed in near/peripersonal space (i.e.

the space within arm-reaching distance) and require fine-grained metric information. On the other hand, allocentric representations have an important role in recognizing objects and scenes, and in planning future movements in far/extrapersonal space (i.e. the space outside arm-reaching distance) (e.g. Berti, Smania, & Allport, 2001; Kosslyn, 1994; Milner & Goodale, 1995; Weiss et al., 2000). According to these considerations, egocentric and allocentric reference systems seem to be functionally related to other theoretical distinctions in the spatial processing domain. One is the distinction between coordinate and categorical spatial processing, that could be related to egocentric and allocentric frames respectively (Jager & Postma, 2003). According to Kosslyn (1987, 1994) categorical representations define invariant non metric relations between objects or between objects and the self (such as above\below, to the left\right of); instead coordinate representations define variable metric relations between objects or between objects and the self (such as the table is 1m far from the door). The other possible distinction is that between near/peripersonal space (i.e. the space where there is direct interaction between objects and body), in which the egocentric frame would predominate, and far/extrapersonal space (i.e. where objects are out of reaching), mainly related to allocentric frames (Berti et al., 2001; Weiss et al., 2000).

The fundamental role played by the frames of reference in several cognitive functions and in action has led to the hypothesis that they may form specialized cognitive mechanisms that rely on specific neural networks. Many studies investigating the cerebral organization of spatial processing in rodents, non-human primates and humans, and many behavioral studies have provided support for the hypothesis of specialization (Bird & Burgess, 2008).

Egocentric and Allocentric Spatial Frames of Reference: Neuroscientific Evidence

Electrophysiological studies in the monkey's brain have shown that parieto-frontal circuits, ventral premotor cortex and posterior parietal cortex are involved in egocentric processing (Cohen & Andersen, 2002; Colby, 1998; Luppino, Murata, Govoni, & Matelli, 1999). In rats, the allocentric processing has been associated to the hippocampus (Espina-Marchant, Pinto-Hamuy, Bustamante, Morales, & Herrera-Marschitz, 2009; Olson & Gettner, 1995). Further, cells with allocentric properties have been individuated in the hippocampal formation of both freely-moving rats (O'Keefe & Dostrovsky, 1971; Taube, Muller, & Ranck, 1990) and monkeys (Rolls & O'Mara, 1995).

To date, only few studies have examined in a direct way the neural networks associated with egocentric and allocentric processing in healthy subjects. In an fMRI study, Vallar et al. (1999) showed that posterior parietal and lateral frontal premotor regions in the right hemisphere are activated more extensively by egocentric information. Galati et al. (2000) confirmed the association of the fronto-parietal network with the egocentric processing, and showed that a subset of these regions was also involved in allocentric tasks. Committeri et al. (2004) compared egocentric and allocentric spatial coding of realistic 3D-objects, and showed that egocentric coding activated mainly areas in the dorsal stream and in frontal lobes, whereas allocentric coding was associated with both dorsal and ventral regions. Zaehle et al. (2007) found a relevant involvement of the precuneus and the medial superior-posterior areas in the processing of egocentric spatial relations, whereas an additional involvement of the right parietal cortex, the ventral visual stream and the hippocampal formation was needed for the allocentric spatial coding.

Some studies investigated the neural bases of several navigational tasks. For example, Rosenbaum and collaborators (Rosenbaum, Ziegler, Winocur, Grady, & Moscovitch, 2004) found a common activation within the medial temporal lobe, in particular the right parahippocampal gyrus. Further, the medial and posterior parietal cortex were associated with egocentric processing, whereas the retrosplenial cortex was associated with allocentric processing. Ghaem et al. (1997) showed the presence of a specific mental navigation network which included the right hippocampus, the left precuneus and the insula. In a PET study, Maguire et al. (1998) found that if participants were involved into a direct mental navigation task the right hippocampus and the right inferior parietal cortex were strongly activated, whereas navigation with detour activated also the left superior and middle frontal gyri.

Although neuropsychological studies with brain-lesioned patients have not addressed directly this topic, they have reported egocentric or allocentric deficits in several disorders. For instance, in optic ataxia (inaccurate visuomotor coordination, Perenin & Vighetto, 1988) and unilateral neglect (failure to explore the contralesional side of space, Vallar, 1998) patients are impaired in perceiving their body mid-sagittal plane, that is a primary egocentric information (Pizzamiglio, Committeri, Galati, & Patria, 2000). Driver (1999) showed that the unilateral neglect may regard the contralesional side of objects independently of the observer's position (object-based neglect). Left "allocentric" neglect seems to be associated with hypoperfusion of right superior temporal gyrus whereas left "egocentric" neglect is more linked to hypoperfusion of the right angular gyrus (Hillis et al., 2005). More recently, Iachini and colleagues (Iachini, Ruggiero, Conson, & Trojano, 2009a) compared left- and right-parietal brain lesioned patients on an egocentric and allocentric spatial memory task. The results showed that right-patients dropped dramatically in egocentric judgments, but performed as well as healthy controls in allocentric judgements. However, left-patients showed a significant impairment in both spatial components. This pattern of results might be interpreted as evidence that the right hemisphere is relatively specialized in processing spatial information according to egocentric frames of reference.

In summary, neuropsychological data show a selective specialization for egocentric and allocentric processing that would be subserved by specific, although partially overlapping, neural areas. In particular, researchers concur on the central role played by the hippocampal (mainly allocentric) and fronto-parietal (mainly egocentric) circuits.

Levels of Attentional Resources: Behavioral Studies

Besides neural bases, egocentric and allocentric frames seem also to differ with regard to the level of attentional resources they require. In 1979, Hasher and Zacks proposed an important framework for research in memory that contrasted automatic and effortful processes. Within this framework they suggested that encoding and retrieving spatial information is so fundamental to the survival of living species that it works almost automatically. Following this suggestion, several studies have investigated the attentional demands of the spatial processes involved in spatial memory. The procedures adopted, however, did not explicitly control the egocentric or allocentric nature of the strategies deployed to solve the tasks and the kind of spatial information processed, although it is possible to make some inference. Tasks that seem to require allocentric processing show no evidence in favor of automatic processes (Andreade & Meudell, 1993; Köhler, Moscovitch, & Melo, 2001; Naveh-Benjamin, 1987, 1988), whereas tasks requiring more egocentric processing highlight the automatic nature of the processes involved (Ellis, 1990; Parkin, Walter, & Hunkin, 1995; Pouliot & Gagnon, 2005). The percentages of accuracy confirm that the two processing systems may differ in terms of attentional demands. For instance, highly accurate performances would be more automatic, while less accurate performances would be more effortful. Accuracy for allocentric tasks usually varies from 37.7% to 56%, whereas for egocentric tasks it ranges from 80 to 90.5% (see Table 1). The fact that the allocentric processing seems to be more effortful is consistent with the idea that it requires more resources to detach from egocentric perspectives.

The Effect of Age on Frames of Reference

The study of the developmental course of egocentric and allocentric spatial processing systems could give useful insights into their characteristics. If it were demonstrated that they are differently affected by aging processes, it would be a further evidence in favor of relatively specialized systems.

So far, age-related changes in basic visuo-spatial abilities, mental imagery and navigational abilities have been investigated (see Iachini, Poderico, Ruggiero, & Iavarone, 2005; Ruggiero, Sergi, & Iachini, 2008). The results obtained are still controversial and it is not yet clear which spatial processes decline with age and which ones are preserved. As regards the egocentric/allocentric distinction, to the best of our knowledge the literature on aging and spatial memory has not directly addressed this issue. In general, several spatial tasks have been used, such as pointing tasks, and the results are interpreted as consistent with the allocentric or the egocentric organization of spatial knowledge. Parkin et al. (1995) used a spatial discrimination task that involved egocentric spatial memory to compare healthy elderly and young people. They found no significant negative effect of age on the spatial performance, but only a slight decline. Instead, Hort et al. (2007) compared healthy elderly, patients with Mild Cognitive Impairment (MCI) and with Alzheimer's disease (AD) on a navigational task that involved either egocentric or allocentric components. Interestingly, they found a significant deficit in the allocentric component in patients but not in healthy elderly people. More recently, Iachini and co-workers (Iachini, Ruggiero, & Ruotolo, 2009b) compared healthy participants aged from 20 to 89 years in a spatial task that required egocentric and allocentric spatial judgments. The results showed that aging had a selective negative impact on the egocentric component starting from the seventies, whereas the allocentric component looked relatively preserved. This suggests that the two components are supported, at least partially, by neural areas that are differently vulnerable to normal aging processes.

Table 1. Percentages of accuracy in previous studies about spatial judgments

	Spatial relations		Task	
	Categorical	Coordinate	Egocentric	Allocentric
Naveh-Benjamin (1988)	+			50%
Ellis (1990)	+		80-90%	
Andreade & Meudell (1993)	+			37.7%
Köhler et al. (2001)		+		56%
Pouliot & Gagnon (2005)	+		90.5%	

OVERVIEW OF THE EXPERIMENTS

The body of neuroscientific and behavioral evidence reviewed in this chapter is consistent with the idea of two specialized systems of spatial processing. Our main purpose is to describe a spatial task devised to measure behaviorally the capacity to use egocentric and allocentric frames of reference and that requires coordinate metric information (Iachini & Ruggiero, 2006).

In the domain of spatial memory, many studies have investigated the influence of several factors on egocentric and allocentric frames of reference. For example, the way of learning spatial information (Presson & Hazelrigg, 1984), the size of spatial layouts (Presson et al., 1989; Roskos-Ewoldsen et al., 1998), the geometric structure of the environment (McNamara et al., 2003) and the degree of familiarity (Evans & Pezdek, 1980; Iachini, Ruotolo, & Ruggiero, 2009c; Ruggiero & Iachini, 2006; Thorndyke & Hayes-Roth, 1982) have been studied. The majority of these studies have adopted pointing tasks that take the following form: "you are (or imagine being) at X, in front of Y, where is Z"? Latency and errors of localization are taken as dependent variables. Overall, the results suggest that familiarity, regularity of the environment and locomotor exploration under multiple perspectives would mitigate orientation-dependent effects and thus favor an allocentric representation. However, it is important to point out that depending on the theoretical definition, different patterns of data may be interpreted as verifying the "allocentric" view. According to the "equiavailable" definition of allocentricity, the position of target-objects should not affect the performance (a flat trend is expected). On the contrary, "environmental" theories point out that the ease of accessing memorized positions depends on whether they are aligned or not with salient axes (e.g. McNamara, 2003). Therefore, a certain degree of ambiguity is intrinsic in these data and this highlights the importance of devising a behavioral task that is able to compare directly both egocentric and allocentric spatial components.

The task we propose here, called "Ego-Allo Task" (EAT), is based on spatial judgments of distance between memorized 3-dimentional real objects previously presented on a desk. Frames of reference are manipulated by asking spatial judgments of relative distance requiring either an egocentric or an allocentric frame. Spatial judgments have been largely adopted to study categorical and coordinate relations (for a review see Jager & Postma, 2003). In general, these studies adopted several "bar and dot" tasks that involved simplistic spatial stimuli presented on papers or computer screens and that required mostly categorical-like judgments (e.g. Bruyer, Scailquin, & Coibion, 1997; Fink et al., 2000; Galati et al., 2000; Sterken, Postma, De Haan, & Dingemans, 1999; Vallar et al., 1999). Instead, our task is clearly coordinate-like as it requires the comparison of subtle metric distances. Egocentric and allocentric frames of reference were manipulated by asking participants to provide spatial judgments either in relation to the viewer ("Which object was closest/farthest to/from you?") or to an object ("Which object was closest/farthest to/from the Cube?"). In humans, egocentric frames of reference represent the primary inter-face between the organism and the environment (e.g. Millar, 1994). For this reason, more processes would be necessary to work out allocentric spatial representations from egocentric representations. This leads to the general hypothesis that the egocentric performance should be more accurate and faster than the allocentric one.

EXPERIMENT 1

In this Experiment we explored the possible influence of near and far spaces on egocentric and allocentric processing of coordinate metric information. A reduced version of this task has already been presented (Iachini & Ruggiero, 2006). We expected that egocentric judgments should be more accurate and faster in near than far space, whereas allocentric judgments should be better in far than near space.

Method

Participants

Forty students (20 males and 20 females) took part in the experiment on a voluntary basis (M_{age}= 25.2 years, SD = 1.72). They were drawn from the Second University of Naples. They understood the instructions without difficulty and none of them were aware of the hypotheses at the time of testing. Further, they were right-handed and had normal or corrected to normal sight.

Setting and Materials

The experiment was carried out in a sound-proofed, comfortable room. Participants sat on a straight-back chair placed centrally at 30 cm from the edge of a desk measuring 100 cm × 150 cm. Six geometrical 3-dimensional objects were chosen like stimuli on the basis of the following criteria: they had simple and elementary geometrical shapes; they were easily recognizable; they could be named without difficulty. These objects were grouped in two series: Series A (Pyramid, Parallelepiped, and Cone) and Series B (Cube, Sphere, and Cylinder). Moreover, these objects could vary on the basis of two physical characteristics: Size and Color. The size comprised two dimensions: Big and Small. Big-objects measured 8 cm × 8 cm except the Parallelepiped and the Cylinder (8 cm × 11 cm), whereas Small-objects measured 6 cm × 6 cm except the Parallelepiped and the Cylinder (6 cm × 9 cm). The color comprised three different tonalities of grey corresponding to dark, medium, and light grey (respectively 75-50-25% of grey). By combining objects, size and color, 9 objects for each series were achieved. For example, the Cone could be small-light or big-dark etc. Still, each series was subdivided in 3 sub-series, respectively A1, A2, and A3; B1, B2, and B3. Each sub-series had a Target-object (T), that is the object with respect to which the allocentric judgments were made. The sub-series A1 comprised: Pyramid-Big-Dark (T), Parallelepiped-Small-Medium, Cone-Small-Light. The sub-series A2 comprised: Pyramid-Small-Medium, Parallelepiped-Big-Light, Cone-Big-Dark (T). The sub-series A3 comprised: Pyramid-Small-Light (T), Parallelepiped-Small-Dark, Cone-Big-Medium. On the other hand, B1 included: Cube-Small-Medium (T), Sphere-Small-Dark, Cylinder-Small-Light; B2 comprised: Cube-Big-Dark, Sphere-Small-Light (T), Cylinder-Small-Medium; B3 included: Cube-Small-Light, Sphere-Big-Medium, Cylinder-Big-Dark (T). These triads were presented according to different spatial arrangements, defined on the basis of three pilot studies, each comprising 15 participants. These studies were carried out in order to: (a) define distinguishable metric distances among objects; (b) determine different levels of metric difficulty; (c) establish the duration of learning time; (d) define the same level of metric difficulty for egocentric and allocentric judgments. The metric difficulty was based on the difference in centimeters between the two distances to be compared; it increased as the difference decreased. For example (see Figure 1), within the sub-series B1, the Cube (T) was 11 cm far from the Sphere and 17 cm from the Cylinder. In this way, the allocentric spatial judgments (e.g. "Which object was closest to the Cube?") were based on a metric difference of 6 cm. On the other hand, the Sphere was 12 cm far from the edge of the cardboard, while the Cylinder was 6 cm; therefore, the egocentric spatial judgments were also based on a metric difference of 6 cm. The six triads of objects (A1, A2, and A3; B1, B2, and B3) were placed centrally on the desk and with respect to the participants' mid-sagittal plane.

To ensure that all triads were presented in the same way for all participants, a cardboard (measuring 50 cm × 30 cm) was used. On this cardboard the shape of the objects was cut out and there was a mark on the lower horizontal side that corresponded to a mark on the desk.

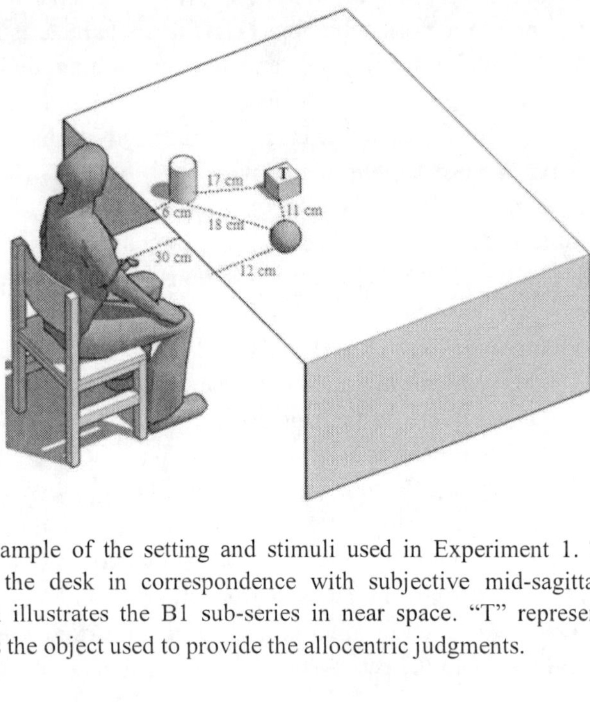

Figure 1. Example of the setting and stimuli used in Experiment 1. The triad was centered on the desk in correspondence with subjective mid-sagittal plane. This configuration illustrates the B1 sub-series in near space. "T" represents the target-object, that is the object used to provide the allocentric judgments.

In the near condition, the cardboards were placed at 30 cm far from the edge of the desk and within the arm-reaching space (50 cm). In the far condition, the cardboards were positioned at 100 cm far from the edge of the desk, that is out of the arm-reaching space.

Procedure

At the beginning of the experiment, participants received written instructions about the task. They were instructed to memorize as accurately as possible the positions and the characteristics of the objects (color and size) that were to be presented. The instructions were then revised orally by the experimenter and a training session started. During training an example of the entire procedure was given by using three common objects (for example: a glass, a cup and a small box). When the procedure was clear, the experimental session began.

Learning phase. At the beginning of each trial, three geometrical 3-dimensional objects were randomly placed on the desk and named. In this way we excluded possible effects due to difficulties in naming the objects. Then the experimenter asked participants to close their eyes and arranged the triad on the desk by means of the proper cardboard. Afterwards, participants had to open their eyes and to memorize the objects and their characteristics. They had to learn each triad while seating still and without turning their heads. The learning phase took 20 sec. Finally, participants had to close their eyes and the experimenter removed the triad. After 5 sec, the testing phase started. This procedure was repeated for all 6 triads.

Testing phase. The experimenter asked participants to give 8 judgments for each triad. There were two egocentric questions: "Which object was closest to you?" and "Which object was farthest from you?"; two allocentric questions: "Which object was closest to the Cube (i.e. Target-object, as shown in Figure 1)?" and "Which object was farthest from the Cube?". In addition, there were four distractors: "Which object was tallest?" and "Which object was largest?"; "Which object was darkest?" and "Which object was lightest?". For each judgment, accuracy and latency were recorded. Latency was recorded by means of a hand-held stopwatch, starting from when the experimenter named the required frame (e.g. Cube) or characteristic until the participant gave the response. Spatial judgments were firstly randomized and then balanced on the basis of a Latin square design. Half of participants started with the series A and half with the series B. Moreover, each participant learned one series in near space and the other in far space. Near/far spaces and the two series were balancced over all participants. At the ending of the testing phase, participants were interviewed to be sure that they had followed the instructions accurately. They were also asked to describe the way in which they performed the task and if there were positions more difficult than others to retrieve. Two participants who did not memorize accurately the stimuli were discarded and replaced. The Experiment was fully completed in approximately 20 minutes.

The experimental design comprised 2 (Egocentric and Allocentric spatial judgments) × 2 (Near and Far spaces) within variables. As dependent variables we had accuracy (mean of correct judgements) and mean latency for correct responses (sec).

Results

Two 2 × 2 ANOVAs (Egocentric/Allocentric and Near/Far) for within subjects designs were used to analyze the data. As regards accuracy, the ANOVA revealed a significant main effect of judgments, $F(1, 39) = 39.99$, $\eta^2 = .51$, $p < .001$, with egocentric judgments ($M = .904$, $SD = .15$) being more accurate than allocentric ones ($M = .73$, $SD = .26$). No significant main effect of Near/Far spaces ($F < 1$) and no interaction ($F < 1$) emerged (see Table 2).

As regards latency, similar results were observed. The ANOVA revealed that egocentric judgments ($M = 1.13$, $SD = 0.43$) were faster than allocentric judgments ($M = 1.82$, $SD = 0.77$), $F(1, 39) = 48.55$, $\eta^2 = .55$, $p < .001$. Neither main effect of Near/Far spaces, $F(1, 39) = 1.20$, $p = .28$, nor interaction ($F < 1$) were significant. As it is common in spatial tasks, we also checked the sample for gender differences. The results did not show significant differences in accuracy, $F(1, 38) = 1.02$, $p = .319$, and latency ($F < 1$).

Overall, the pattern of data shows that the task based on spatial judgments of distance is able to distinguish between egocentric and allocentric processing of metric information. A strong facilitation of the egocentric judgments over the allocentric ones in accuracy and latency emerged. Percentages of accuracy were quite similar to previous studies, respectively 90% vs 73%. In line with our hypothesis, the results confirm that the allocentric processing is more difficult than the egocentric processing. The strength of the effect sizes confirmed the robustness of the results and consequently the discriminatory power of the task in distinguishing egocentric and allocentric processing. In contrast, the manipulation of the distance between objects and the observer (near/far spaces) did not affect the performance. Probably, this result is due to the low ecological impact of the experimental manipulation that operationalized the two spaces. Indeed, all stimuli were easily visible and participants were seated throughout the experimental session. However, before taking for granted the reliability of the task, a criticism must be faced: the results could have been determined by an experimental artefact that facilitated the egocentric coding. More precisely, the position of the egocentric frame was always stable as participants seated still throughout the experimental session, while the positions of the allocentric targets were different for each triad. This variability of the allocentric positions could have undermined the allocentric performance whereas the stability of the body could have favored the egocentric performance. Experiment 2 was aimed at controlling this spurious factor.

EXPERIMENT 2

In order to eliminate the possible facilitating effect due to the stable egocentric position, in this experiment each triad was studied from a different learning position. In this way, both egocentric and allocentric frames of reference varied for each configuration. Further, stimuli were all presented in near space. Two conditions were compared named "Stable" (the same as Experiment 1) and "Variable" in which participants had to move with respect to each triad of objects. If the stability of the egocentric position facilitates the performance, then the egocentric advantage should disappear in the "Variable" condition. This hypothesis should be verified by a significant interaction.

Method

Participants

Forty students (20 males and 20 females; M_{age} = 24.97 years, SD = 2.69) from the Second University of Naples were recruited on a voluntary basis and randomly assigned to the "Stable" or "Variable" conditions. They were right-handed and had normal or corrected to normal sight.

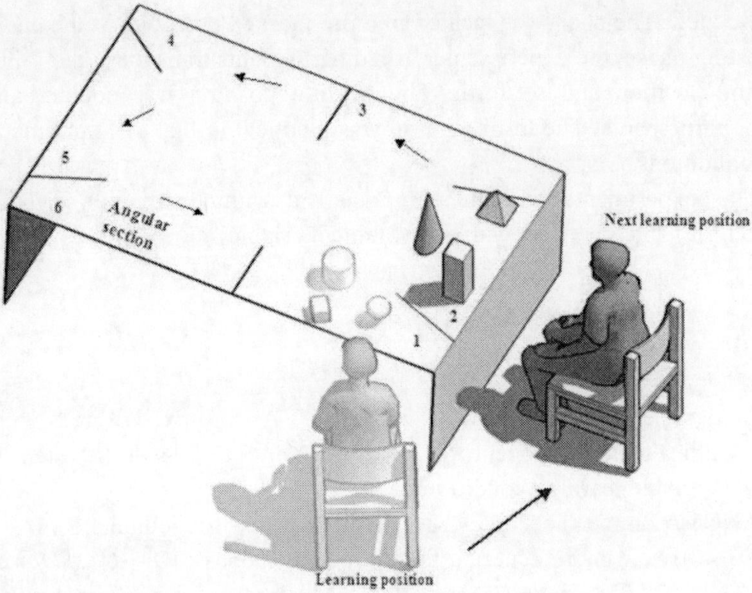

Figure 2. Example of the setting and stimuli used in the Variable condition of Experiment 2. The numbers from 1 to 6 indicate the angular sections worked out on the desk. Participants had to seat in front of the first angular section randomly assigned (for example, the section n° 1) and from this position they had to move to the next position in a counter-clockwise order, as illustrated by the black arrows.

Setting and Materials

As regards the "Stable" condition, setting and materials were identical to those of Experiment 1 in near space. As shown in Figure 2, in the "Variable" condition six different sections (each one of 60 degrees) were worked out on the desk (invisible for participants) and each triad was placed on one of these sections. Participants had to study the six triads of objects by seating in front of the assigned section. In this way, both egocentric and allocentric frames of reference varied for each configuration. The order of learning positions was counterbalanced across participants.

Procedure

The "Stable" group followed the same procedure as in Experiment 1. In the "Variable" group, participants sat in front of the desk in correspondence of each section. The chair was located in a pre-marked position. At the ending of the testing phase, the experimenter asked participants to move to the following learning position and so forth. The starting position was counterbalanced within participants. The testing phase was identical to that of Experiment 1 in both conditions.

The experimental design comprised a within variable (Egocentric/ Allocentric) and a between variable (Stable/Variable).

Results

Analyses were based on two 2-way ANOVAs for mixed designs with a 2-level within factor (Egocentric/Allocentric) and a 2-level between factor (Stable/Variable learning conditions).

ANOVA on mean accuracy revealed that egocentric judgments ($M = .901$, $SD = .146$) were more accurate than allocentric ones ($M = .69$, $SD = .199$), $F(1, 38) = 28.514$, $\eta^2 = .42$, $p < .001$. No significant difference emerged between Stable and Variable learning conditions ($F < 1$) and no significant interaction ($F < 1$; see Table 2).

ANOVA on latency confirmed that egocentric judgments were faster than allocentric ones, $F(1, 38) = 31.084$, $\eta^2 = .45$, $p < .001$. The related means were: egocentric = 1.28, $SD = 0.445$; allocentric = 1.88, $SD = 0.814$. Further, the Stable ($M = 1.34$, $SD = 0.418$) was significantly faster than the Variable condition ($M = 1.82$, $SD = 0.716$), $F(1, 38) = 9.122$, $\eta^2 = .20$, $p < .005$. Finally, no significant interaction was found, $F < 1$. As in the previous experiment, no gender differences appeared on both accuracy, $F < 1$, and latency, $F(1, 38) = 1.441$, $p = .24$.

Table 2. Percentages of accuracy in the Ego-Allo Task (EAT)

	Task	
Experimental conditions	Egocentric	Allocentric
	Experiment 1	
Near space	91%	72.5%
Far space	90%	74.2%
	Experiment 2	
Stable	93%	73%
Variable	88%	65%

The results confirm the reliability of the task as the egocentric advantage in accuracy and latency is not disrupted when the egocentric learning position varies for each triad of objects. However, the data show that the judgments are provided more slowly in the variable than in the stable condition. It is likely that more time is needed to update the spatial relation between the observer and external frames of reference, such as the desk and the walls (see McNamara, 2003). Even though the variable spatial relationship between environment and observer lengthened global processing time, it did not alter the egocentric facilitation. Again, the strength of the effect sizes confirmed the robustness of the data and the clear egocentric primacy over the allocentric processing.

CONCLUSION

The principal goal of the experiments reported here was to compare directly memories for egocentric and allocentric processing of coordinate metric relations in an experimentally controlled situation that was as closer as possible to everyday situations. The EAT involved real 3-dimensional stimuli that were displaced on a desk. The procedure was simple as required learning and then retrieving spatial relations between the stimuli and the subject (egocentric judgments) or another object (allocentric judgments). The task was ease to administer and behavioral responses were recorded with no aid of technical devices. For all these reasons, this task could be adopted with different populations such as children, elderly people (Iachini et al., 2009b) and patients suffering from brain-damage (Iachini et al., 2009a).

The results we obtained confirm the good discriminatory power of the task. Indeed, in the two experiments the difference between allocentric and egocentric judgments was always strong and always in favor of egocentric judgments. This robust effect was supported by high effect sizes and was not influenced by artifactual factors.

Experiment 1 showed a clear advantage of the egocentric processing over the allocentric one in terms of both accuracy and latency. This advantage was not altered by the distance between individuals and objects, that is the spaces of coding within or outside arm-reaching. Probably, this is due to the capacity of normal healthy adults to process spatial information equally well in different spaces of coding. On the other hand, it could also be due to a low ecological relevance of the experimental manipulation: in all cases participants learned spatial information while seated and stimuli displayed in "far space" were easily perceivable (Iachini et al., 2009b). The question of the processing of egocentric and allocentric information in near and far spaces has been recently addressed by Iachini and colleagues (Iachini et al., 2009a). In their study, left and right brain-lesioned patients showed selective impairments in egocentric and allocentric processing. More specifically, the left hemisphere patients revealed difficulties in processing both egocentric and allocentric information, especially in far space. Instead, the right hemisphere patients were specifically impaired in egocentric judgments and this effect was particularly strong in near space. These observations would suggest that different categories of patients might be selectively affected by the range of space.

Experiment 2 discarded the possibility that the results were due to spurious factors, such as the stability vs the variability of the frames of reference. Even though a general slowing was observed when the learning positions were variable, this could be ascribed to the spatial processes needed to update the relationship between the observer and external allocentric cues. However, these processes did not alter the egocentric advantage over the allocentric component.

Overall, the two Experiments confirm the primacy of the egocentric frames which represent the primary inter-face between the organism (the viewer's body) and the environment (Millar, 1994). In clarifying some aspects of the perception\action model, Milner and Goodale (2008) hypothesize a close link between the egocentric and allocentric processing and the functions of visual streams. The model proposes a distinction between two visual streams: a dorsal stream that processes information useful to control action, and a ventral stream that processes information useful to recognize objects. The ventral stream should transform visual information into allocentric perceptual representations, while the dorsal stream should use on-line information about the egocentric organization of objects. Interestingly, Sterken et al. (1999) suggested that the egocentric system is specialized in processing information to guide motor behavior by analyzing the metric properties of spatial relations. In contrast, spatial representations allocentrically specified are needed to plan trajectories and to recognize objects (Sterken et al., 1999; see also Kosslyn, 1994).

According to an evolutionary framework, the capacity to perceive and represent spatial information is affected by the adaptive specializations of the species in their interaction with the environment. In a seminal comparative experiment, Delius and Hollard (1995) presented rotated stimuli to human beings and pigeons. The results revealed that humans were affected by the degree of rotation, whereas pigeons recognized stimuli independently of the variance of rotation. Since humans assumed the erected position, spatial information has been acquired frontally, that is egocentrically. Instead, pigeons perceive spatial relations horizontally from a survey view. For this reason, they are able to capture easily the invariant allocentric characteristics of spatial information. On contrast, as suggested by Millar (1994), in humans more processes are needed to detach from the egocentric perspective in order to represent spatial information allocentrically and this computational effort should be paid in terms of errors and latency.

Coming back to our results, they could provide a potential contribution to the literature about automatic vs effortful encoding of spatial information. According to Hasher and Zacks (1979), the automatic and effortful distinction is not a dichotomous one, rather encoding processes are situated along a continuum of attentional demands. As suggested by Ellis (1990) and, more recently, by Pouliot and Gagnon (2005), the percentage of accuracy can be taken as an index of the rather automatic nature of a process. In the present experiments the percentages of accuracy in the egocentric processing are extremely high (about 90%), whereas in the allocentric processing they range from 69% to 73%. As shown in Table 2, they are similar to the percentages reported in previous studies where egocentric accuracy ranged from 80% to 90.5% and allocentric accuracy varied from 37% to 56% (see Table 1). However, we must be cautious in comparing our data with previous ones given relevant differences. For instance, in previous studies there was no direct comparison between egocentric and allocentric processing and the nature of the spatial tasks was not clarified (see Pouliot & Gagnon, 2005). Further, categorical-like tasks (i.e. right/left of ...) requiring the relocation or the discrimination of pictures, drawings and so forth were adopted (Andreade & Meudell, 1993; Ellis, 1990; Naveh-Benjamin, 1988; Pouliot & Gagnon, 2005). Instead, our task involves coordinate spatial relations, compares directly the egocentric and allocentric frames of reference and the amount of metric difficulty is the same for both judgments. This last factor is crucial if we want to be sure that differences in performance are due to intrinsic characteristics of egocentric and allocentric processing and not to the presence of spurious factors. At this stage, we can accept the idea that when the task requires spatial judgements and involves real 3-D objects there is a facilitation for egocentric rather than allocentric processing. Further studies would be necessary to disentangle whether this advantage is due to attentional factors or to a mere spatial component.

In conclusion, in this chapter we reviewed several lines of evidence that converge on the idea that egocentric and allocentric spatial processing may represent specialized systems. It can be speculated that this selective specialization results from evolutionary pressures. The necessity to perform and control actions and to recognize objects and scenes represent two primary adaptive functions. These functions may have led to the specialization of dedicated systems in neighboring areas of the brain according to a "snowball mechanism" as suggested by Kosslyn (1987). Egocentric and allocentric components may represent two key processes in the service of the primary functions of action and recognition.

References

Andrade, J. & Meudell, P. (1993). Is spatial information encoded automatically? *The Quarterly Journal of Experimental Psychology, 46,* 365-375.

Avraamides, M. & Kelly, J. (2008). Multiple systems of spatial memory and action. *Cognitive processing, 9(2)*, 93-106.

Berti, A., Smania, N. & Allport, A. (2001). Coding of far and near space in neglect patients. *Neuroimage, 14*, 98-102.

Bird, C. M. & Burgess, N. (2008). The hippocampus and memory: insights from spatial processing. *Nature Reviews Neuroscience, 9(3)*, 182-194.

Bruyer, R., Scailquin, J. C. & Coibion, P. (1997). Dissociation between categorical and coordinate spatial computations: modulation by cerebral hemispheres, task properties, mode of response, and age. *Brain and Cognition, 33*, 245-277.

Cohen, Y. E. & Andersen, R. A. (2002). A common reference frame for movement plans in the posterior parietal cortex. *Nature Review Neuroscience, 3*, 553-562.

Colby, C. L. (1998). Action-oriented spatial reference frames in cortex. *Neuron, 20*, 15-24.

Committeri, G., Galati, G., Paradis, A., Pizzamiglio, L., Berthoz, A. & Le Bihan, D. (2004). Reference frame for spatial cognition: different brain areas are involved in viewer-, objects-, and landmark-centered judgments about objects location. *Journal of Cognitive Neuroscience, 9*, 1517-1535.

Delius, J. D. & Hollard, V. D. (1995). Orientation invariant pattern recognition by pigeons (Columba livia) and humans (Homo sapiens). *Journal of Comparative Psychology, 109*, 278-290.

Diwadkar, V. A. & McNamara, T. P. (1997). Viewpoint dependence in scene recognition. *Psychological Science, 8(4)*, 302-307.

Driver, J. (1999). Egocentric and object-based visual neglect. In: N., Burgess, K. J. Jeffery, & J. O'Keefe (Eds.), *Spatial functions of the hippocampal formation and the parietal cortex* (pp. 67-89). Oxford: Oxford University Press.

Eilan, N. (1993). Molyneux's question and the idea of an external world. In N. Eilan, R. A. McCarthy, & B. Brewer (Eds.), *Spatial representation: Problems in philosophy and psychology* (pp. 236-255). Malden, MA, US: Blackwell Publishing.

Ellis, N. R. (1990). Is memory for spatial location automatically encoded? *Memory and Cognition, 18*, 584-592.

Espina-Marchant, P., Pinto-Hamuy, T., Bustamante, D., Morales, P. & Herrera-Marschitz, M. (2009). Rat strain influences the use of egocentric learning strategies mediated by neostriatum. *Experimental Brain Research, 193*, 205-212.

Evans, G. W. & Pezdek, K. (1980). Cognitive mapping: knowledge of real-world distance and location. *Journal of Experimental Psychology: Learning, Memory, and Cognition, 6*, 13-24.

Fink, G. R., Marshall, J. C., Shah, N. J., Weiss, P. H., Halligan, P. W., Grosse-Ruyken, M., Ziemons, K., Zilles, K. & Freund, H. J. (2000). Line bisection judgements implicate right parietal cortex and cerebellum as assessed by fMRI. *Neurology, 54*, 1324-1331.

Franklin, N. & Tversky, B. (1990). Searching imagined environments. *Journal of Experimental Psychology: General, 119*, 63-76.

Ghaem, O., Mellet, E., Crivello, F., Tzourio, N., Mazoyer, B., Berthoz, A. & Denis, M. (1997). Mental navigation along memorized routes activates the hippocampus, precuneus, and insula. *Neuroreport, 8,* 739-744.

Galati, G., Lobel, E., Vallar, G., Berthoz, A., Pizzamiglio, L. & Le Bihan, D. (2000). The neural basis of egocentric and allocentric coding of space in humans: a functional magnetic resonance study. *Experimental Brain Research, 133,* 156-164.

Hasher, L. & Zacks, R. T. (1979). Automatic and effortful processes in memory. *Journal of Experimental Psychology: General, 108*, 356-373.

Hillis, A. E., Newhart, M., Heidler, J., Barker, P. B., Herskovits, E. H. & Degaonkar M. (2005). Anatomy of spatial attention: insights from perfusion imaging and hemispatial neglect in acute stroke. *The Journal of Neuroscience, 25*, 3161-3167.

Hort, J., Laczó, J., Vyhnálek, M., Bojar, M., Bureš, J. & Vlček, K. (2007). Spatial navigation deficit in amnestic mild cognitive impairment. *Proceedings of the National Academy of Sciences of the United States of America, 104,* 4042-4047.

Humphreys, G. W. & Riddoch, M. J. (1994). Attention within-object and between-object spatial representations: Multiple sites for visual selection. *Cognitive Neuropsychology, 11*, 207-241.

Iachini, T., Poderico, C., Ruggiero, G. & Iavarone, A. (2005). Age differences in mental scanning of locomotor maps. *Disability and rehabilitation, 27,* 741-752.

Iachini, T. & Ruggiero, G. (2006). Egocentric and allocentric spatial frames of reference: a direct measure. *Cognitive Processing, 7*, 126-127.

Iachini, T., Ruggiero, G., Conson, M. & Trojano, L. (2009a). Lateralization of egocentric and allocentric spatial processing after parietal brain lesions. *Brain and cognition, 69,* 514-520.

Iachini, T., Ruggiero, G. & Ruotolo, F. (2009b). The effect of age on egocentric and allocentric spatial frames of reference. *Cognitive Processing, 10,* 222-224.

Iachini, T., Ruotolo, F. & Ruggiero, G. (2009c). The effects of familiarity and gender on spatial representation. *Journal of Environmental Psychology, 29*, 227-234.

Jager, G. & Postma, A. (2003). On the hemispheric specialization for categorical and coordinate spatial relations: a review of the current evidence. *Neuropsychologia, 41,* 504-515.

Köhler, S., Moscovitch, M. & Melo, B. (2001). Episodic memory for object location versus episodic memory for object identity: do they rely on distinct encoding processes? *Memory and Cognition, 29*, 948-959.

Kosslyn, S. M. (1987). Seeing and imagining in the cerebral hemispheres: A computational analysis. *Psychological Review*, *94*, 148-175.

Kosslyn, S. M. (1994). *Image and brain: the resolution of the imagery debate*. Cambridge, MA: MIT Press.

Levinson, S. C. (1996). Frames of reference and Molyneaux's question: Crosslinguistic evidence. In P. Bloom, M. A. Peterson, L. Nadel, & M. F. Garrett (Eds.), *Language and space* (pp. 109-169). Cambridge, MA: MIT Press.

Luppino, G., Murata, A., Govoni, P. & Matelli, M. (1999). Largely segregated parieto-frontal connections linking rostral intraparietal cortex (areas AIP and VIP) and the ventral premotor cortex (areas F5 and F4). *Experimental Brain Research, 128*, 181-187.

Maguire, E. A., Burgess, N., Donnet, J. G., Frackowiak, R.S.J., Frith, C. D. & O'Keefe J. (1998). Knowing where and getting there: a human navigation network. *Science, 280,* 921-924.

McNamara, T. P. (2003). How are the locations of objects in the environment represented in memory? In C. Freksa, W. Brauer, C. Habel, & K. Wender (Eds.), *Spatial cognition III: Routes and navigation, human memory and learning, spatial representation and spatial reasoning* (pp. 174-191). Berlin: Springer-Verlag.

McNamara, T. P., Rump, B. & Werner, S. (2003). Egocentric and geocentric frames of reference in memory of large-scale space. *Psychonomic Bulletin & Review, 10*, 589-595.

Millar, S. (1994). *Understanding and Representing Space. Theory and Evidence from Studies with Blind and Sighted Children*. Oxford: Clarendon Press.

Milner, A. D. & Goodale, M. A. (1995). *The visual brain in action.* Oxford: Oxford University Press.

Milner, A. D. & Goodale, M. A. (2008). Two visual systems re-viewed. *Neuropsychologia, 46,* 774-785.

Mou, W. & McNamara, T. P. (2002). Intrinsic frames of reference in spatial memory. *Journal of Experimental Psychology: Learning, Memory, and Cognition, 28*, 162-170.

Naveh-Benjamin, M. (1987). Coding of spatial location information: an automatic process? *Journal of Experimental Psychology: Learning, Memory, and Cognition, 13,* 595-605.

Naveh-Benjamin, M. (1988). Recognition memory of spatial location information: Another failure to support automaticity. *Memory and Cognition, 16,* 437-445.

O'Keefe, J. & Dostrovsky, J. (1971). The hippocampus as a spatial map. Preliminary evidence from unit activity in the freely-moving rat. *Brain Research, 34*, 171-175.

O'Keefe, J. & Nadel, L. (1978). *The hippocampus as a cognitive map.* Oxford: Clarendon Press.

Olson, C. R. & Gettner, S. N. (1995). Object-centered direction selectivity in the macaque supplementary eye field. *Science, 269*, 985-988.

Paillard, J. (1991). Motor and representational framing of space. In J. Paillard (Ed.), *Brain and Space* (pp. 163-182). Oxford, New York, Tokyo: Oxford University Press.

Pani, J. R. & Dupree, D. (1994). Spatial reference systems in the comprehension of rotational motion. *Perception, 23*, 929-946.

Parkin, A. J., Walter, B. M. & Hunkin, N. M. (1995). Relationships between normal aging, frontal lobe function, and memory for temporal and spatial information. *Neuropsychology, 9*, 304-312.

Perenin, M. T. & Vighetto, A. (1988). Optic ataxia: a specific disruption in visuomotor mechanisms. Different aspects of the deficit in reaching for objects. *Brain, 111*, 643-674.

Piaget, J. & Inhelder, B. (1967). *The child's conception of space*. New York: Norton.

Pinker, S. (1984). Visual cognition: an introduction. *Cognition, 18(1-3),* 1-63.

Pizzamiglio, L., Committeri, G., Galati, G. & Patria, F. (2000). Psychophysical properties of line bisection and body midline perception in unilateral neglect. *Cortex, 36*, 469-484.

Pouliot, S. & Gagnon, S. (2005). Is egocentric space automatically encoded? *Acta Psychologica, 118,* 193-210.

Presson, C. C., DeLange, N. & Hazelrigg, M. D. (1989). Orientation specificity in spatial memory: What makes a path different from a map of a path? *Journal of Experimental Psychology: Learning, Memory, and Cognition, 15*, 887-897.

Presson, C. C. & Hazelrigg, M. D. (1984). Building spatial representations through primary and secondary learning. *Journal of Experimental Psychology: Learning, Memory, and Cognition, 10*, 716-722.

Rieser, J. J. (1989). Access to knowledge of spatial structure at novel points of observation. *Journal of Experimental Psychology: Learning, Memory, and Cognition, 15*, 1157-1165.

Rolls, E. T. & O'Mara, S. M. (1995). View-responsive neurons in the primate hippocampal complex. *Hippocampus, 5*, 409-424.

Rosenbaum, R.S., Ziegler, M., Winocur, G., Grady, C.L. & Moscovitch, M. (2004). "I have often walked down this street before": fMRI studies on the hippocampus and other structures during mental navigation of an old environment. *Hippocampus, 14,* 826-835.

Roskos-Ewoldsen, B., McNamara, T. P., Shelton, A. L. & Carr, W. (1998). Mental representations of large and small spatial layouts are orientation dependent. *Journal of Experimental Psychology: Learning, Memory, and Cognition, 24*, 215-226.

Ruggiero, G. & Iachini, T. (2006). The effect of familiarity on egocentred and allocentred spatial representations of the environment. *Cognitive Processing, 7*, 88-89.

Ruggiero, G., Sergi, I. & Iachini, T. (2008). Gender differences in remembering and inferring spatial distances. *Memory, 16* (*8*), 821-835.

Schenk, T. (2006). An allocentric rather than perceptual deficit in patient D. F. *Nature Neuroscience, 9* (*11*), 1369-1370.

Shelton, A. L. & McNamara, T. P. (1997). Multiple views of spatial memory. *Psychonomic Bulletin & Review, 4,* 102-106.

Sterken, Y., Postma, A., De Haan, E. H. F. & Dingemans, A. (1999). Egocentric and exocentric spatial judgements of visual displacement. *Quarterly Journal of Experimental Psychology, 52*, 1047- 1055.

Taube, J. S., Muller, R. U. & Ranck, J. B. Jr. (1990). Head-direction cells recorded from the postsubiculum in freely moving rats. II. Effects of environmental manipulations. *Journal of Neuroscience, 10*, 436-447.

Thorndyke, P. W. & Hayes-Roth, B. (1982). Differences in spatial knowledge acquired from maps and navigation. *Cognitive Psychology, 14*, 560-589.

Vallar, G. (1998). Spatial hemineglect in humans. *Trends in Cognitive Sciences, 2*, 87-97.

Vallar, G., Lobel, E., Galati, G., Berthoz, A., Pizzamiglio, L. & Le Bihan, D. (1999). A fronto-parietal system for computing the egocentric spatial frame of reference in humans. *Experimental Brain Research, 124*, 281-286.

Waller, D., Montello, D. R., Richardson, A. E. & Hegarthy, M. (2002). Orientation specificity and spatial updating of memories for layouts. *Journal of Experimental Psychology: Learning, Memory, and Cognition, 28*, 1051-1063.

Weiss, P. H., Marshall, J. C., Wunderlich, G., Tellmann, L., Halligan P. W., Freund, H. J., Zilles, K. & Fink, G. R. (2000). Neural consequences of acting in near versus far space: a psychological basis for clinical dissociations. *Brain, 123*, 2531-2541.

Zaehle, T., Jordan, K., Wüstenberg, T., Baudewig, J., Dechent, P. & Mast, F.W. (2007). The neural basis of the egocentric and allocentric spatial frame of reference. *Brain Research, 1137(1),* 92-103.

In: Spatial Memory
Editor: Jacob B. Thomas, pp. 77-99
ISBN: 978-1-61668-139-5

Chapter 3

ALLOCENTRIC SPATIAL MEMORY IN YOUNG AND OLDER ADULTS: A COMPARISON OF TWO-DIMENSIONAL MAZE LEARNING AND OBJECT-LOCATION MEMORY

Roy P.C. Kessels[1,2*], Sofie Geurts[3], Paul A.T.M. Eling[1] and Helmut Hildebrandt[4,5]

[1]Donders Institute for Brain, Cognition and Behaviour, Radboud University Nijmegen, The Netherlands

[2]Departments of Medical Psychology and Geriatric Medicine, Radboud University Nijmegen Medical Centre, The Netherlands

[3]Department of Clinical Psychology, Canisius Wilhelmina Hospital, Nijmegen, The Netherlands

[4]Department of Neurology, Municipal Hospital of Bremen, Germany

[5]Department of Psychology, University of Oldenburg, Germany

ABSTRACT

There is abundant evidence that allocentric spatial memory declines with advancing age. However, spatial memory is not a unitary construct,

* Corresponding author: Radboud University Nijmegen, Donders Institute for Brain, Cognition and Behaviour, Centre for Cognition, PO Box 9104, 6500 HE Nijmegen, The Netherlands, Phone: +3124 3612541, Fax: +3124 3616066, Email: r.kessels@donders.ru.nl

but can be subdivided into memory for object locations and maze learning. Furthermore, performance on these functions may also be affected by age-associated decline in working memory, executive functioning (notably visuospatial planning) and long-term memory in general. It was the aim of the current study to investigate these aspects in combination in a group of younger (N = 25) and older (N = 25) adults. Maze learning was measured with a computer adaptation of the Morris Water Maze. Object-location memory function was measured using the Object Relocation paradigm, also including task conditions assessing object-location binding and positional memory. Furthermore, spatial working memory was measured with the Spatial Span subtest of the WMS-III, visual long-term memory with the Doors Test and visuospatial planning with the Key Search test. The results clearly showed large (d > 0.8) age effects on all aspects of object-location memory, spatial working memory and visual long-term memory, but not on visuospatial planning and maze learning. Also, spatial working-memory decline was highly correlated with the performance on all other tasks. These results suggest that spatial memory and spatial learning are differentially affected by age and may thus rely on distinct neurocognitive processes.

Keywords: maze learning; spatial memory; aging; working memory; executive function; allocentric

INTRODUCTION

Spatial learning and memory are aspects of cognition that are highly relevant for every-day life functioning. They help us to orient ourselves in place, to navigate and to localise and retrieve objects around us. Also, spatial memory problems are among the most frequent complaints in older age or in patients visiting a memory clinic. Several aspects of spatial learning and memory can be distinguished. Firstly, it entails the learning of routes or paths in our environment. Secondly, it requires the ability to remember positional information and to associate these with objects (i.e., object-location memory). Route or path learning has been studied extensively using rodent models. Typically, a rat or mouse learns a route within a maze in order to reach an object, followed by a positive reward (Yamada, Kawai, Iwasaki, & Ichitani, 2002). The Morris Water Maze (MWM) has been used frequently in this respect. It consists of a round swimming pool containing a hidden platform. The rat or mouse is placed inside the pool and has to find the hidden platform in order to get out of the water. Objects placed outside the swimming area can be used by the animal as external cues to orient itself and to locate the platform sooner, despite different starting points (Vorhees & Williams, 2006). With respect to the neuroanatomical correlates, there is abundant evidence that in rodents the hippocampus is critically involved; lesions of this structure result in impaired maze learning (Clark, Broadbent, & Squire, 2007).

Maze learning has been studied in humans too, using virtual-reality tasks, indicating an important role not only for the hippocampus (Feigenbaum & Morris, 2004), but also for the parietal cortex (Weniger, Ruhleder, Wolf, Lange, & Irle, 2009). In order to find a fixed location from different perspectives, such as in the two-dimensional MWM, an allocentric representation is required. That is, one has to have a viewer-independent frame of reference containing external reference points. This allocentric frame of reference can be regarded as a cognitive map, in which spatial relations between objects are represented, and that is predominantly subserved by the hippocampus (Mou, McNamara, Rump & Xiao, 2006; O'Keefe & Nadel, 1978). In contrast, an egocentric representation consists of a viewpoint-dependent frame of reference, in which the spatial relations between the viewer and the objects are relevant; this reference frame relies on the posterior parietal cortex (Zaehle, Jordan, Wüstenberg, Baudewig, Dechent & Mast, 2006; Rogers & Kesner, 2006; Weniger et al., 2008). In addition, the prefrontal cortex also plays an important role in the search aspect of maze learning (Yoon, Okada, Jung, & Kim, 2008; Karnath & Wallesch, 1992).

With respect to human analogues of the water maze paradigm, different approaches have been adopted. Many researchers use small-space two-dimensional computer paradigms, in which locations have to be searched or remembered within a circle from different perspectives, either with or without external cues present, but large scale real-life maze learning has also been examined (see Fitting, Allen & Wedell, 2007, for an overview). Since the performance on small-scale spatial memory tasks has been found to generalize to the performance in actual large-scale maze learning (Fitting et al., 2007), the present study focuses on age-related differences on two-dimensional small-scale virtual maze learning paradigms that can be controlled better.

Studies on two-dimensional virtual maze learning in older adults showed a lowered performance compared to young adults (Driscoll et al., 2004; Moffat & Resnick, 2002). However, there are some important methodological limitations. For example, in the studies by Driscoll et al. (2004) and Moffat and Resnick (2002) participants were informed that the hidden platform had a fixed location, limiting the emphasis on spatial search that is an important aspect in the original animal maze-learning paradigm. Consequently, the task versions of Driscoll et al. (2004) and Moffat and Resnick (2002) predominantly tapped object-location memory, that is, memory for a fixed location, rather than active spatial search. The present study extends these findings by examining age-related performance on a two-dimensional virtual maze learning task in which participants are instructed to find a hidden platform and subsequently find out by trial and error that this was always located at a fixed position. This is more in agreement with the trial and error learning in the original animal version of the MWM.

Next to maze learning (or spatial search), another important aspect of spatial memory is the ability to store and remember object locations without navigation. There is clear evidence that object-location memory is not a unitary process, but entails separate cognitive operations (Kessels, Kappelle, De Haan & Postma, 2002). One process that can be distinguished is memory for fine-grained locations, that is, memory for coordinate information. This operation relies upon the right hippocampus (Kessels, Hendriks, Schouten, Van Asselen & Postma, 2004) and the right parietal lobe (Van Asselen, Kessels, Kappelle & Postma, 2008). A second cognitive process that is important for object-location memory is the binding of objects to locations. Neuroimaging studies have shown involvement of the hippocampus in spatial binding (Piekema, Kessels, Mars, Peterson & Fernández, 2006) and lesion studies have demonstrated that hippocampal lesions may impair the ability to bind objects to locations (Chalfonte, Verfaellie, Johnson & Reiss, 1996; Kessels et al., 2004). In addition, object location binding may also involve processing of categorical relations between objects, for example, remembering that the chair is located to the left of the table. This may be mediated by the left parietal cortex (Brown & Kosslyn, 1993). Thus, two important aspects of spatial memory can be distinguished: route or maze learning on the one hand and object-location memory on the other hand. These can be selectively impaired following brain damage (Postma, Van Asselen & Kessels, 2008; D'Hooge & De Deyn, 2001).

Since spatial memory is not a unitary construct, and both prefrontal, parietal and hippocampal brain regions are differentially involved in the various aspects of spatial-memory function, an important question is whether differential aging effects may be found for these spatial components. In rodents, there is evidence that specifically allocentric spatial memory deteriorates, whereas egocentric representations remain relatively intact (Begega, Cienfuegos, Rubio, Santín, Miranda & Arias, 2001). This may be due to age-related neural changes in the hippocampus (Murphy, Rahnama & Silva, 2005). In humans, Driscoll, Hamilton, Yeo, Brooks and Sutherland

(2005) have examined three age groups on a virtual Morris water maze (vMWM) and found that older people had longer search paths than younger adults. A study by Moffat and Resnick (2002) demonstrated that older people needed more time than younger adults to reach the platform in a vMWM, which may have been the result of inefficient spatial planning. However, older adults also performed worse on recall of the location of the platform (as assessed by having the participants to draw a map of the maze and the platform's location in it), regardless of search strategy. Both studies did not find any age effects on the condition in which the platform was visible, suggesting that it may be a combination of object-location memory and spatial planning that deteriorates with aging.

Various studies examining aging in relation to object-location memory have shown age-related decline. For example, Perlmutter, Metzger, Nezworski and Miller (1981) examined young and older participants and found impaired spatial memory in older adults compared to young adults. Moreover, age-related memory decline was reported in explicit retrieval of the locations of objects presented in a virtual room (Kessels, Te Boekhorst, & Postma, 2005), and large age differences were also found on both object-location memory and positional memory (Kessels, Hobbel & Postma, 2007). Finally, Cooney and Arbuckle (1997) showed age-related deterioration of spatial-location memory and Cherry, Park and Donaldson (1993) demonstrated impaired memory for three-dimensional spatial information in older adults.

While the aforementioned studies provide clear evidence for an age-related decline in both maze learning and object-location memory, no study has examined the combination of these spatial memory functions, allowing for a direct comparison of the age effects on these types of processing. Furthermore, it is unclear whether this age-related spatial-memory decline is merely the result of a general age-associated cognitive decline. For example, object-location memory does not only rely on memory for spatial information, but also depends on (visual) recognition memory in general (Postma, Kessels, & Van Asselen, 2004), a cognitive function that has also been found to deteriorate due to normal aging (Seo, Lee, Choo, Youn, Kim, Jhoo, Suh, Paek, Jun & Woo, 2007; Fjell, Walhovd, Reinvang, Lundervold, Dale, Quinn, Makris & Fischl, 2005). In turn, maze learning depends on spatial working memory (i.e., keeping track of locations that have already been visited and updating them) and spatial planning ability. Cornoldi et al. (2007) showed that older adults performed worse than younger adults on measures of spatial working memory, due to a heightened susceptibility for intrusion errors (i.e., selecting previously visited locations). Additionally, there is evidence that prefrontal "working-memory" regions interact with the parietal cortex (Wager & Smith, 2003) and the hippocampus (Piekema et al., 2006; Piekema, Fernández, Postma, Hendriks, Wester & Kessels, 2007) during spatial working memory tasks, areas that are also implicated in object-location memory and maze learning. Finally, spatial planning ability as assessed by setting out a path on a map using a set of rules has been found to deteriorate in older adults (Allain, Nicoleau, Pinon, Etcharry-Bouyx, Barré, Dubas & Le Gall, 2005).

The present study is the first to examine aging effects on virtual maze learning and object-location memory in combination, taking working and recognition memory function as well as spatial search into account. We examined these functions in groups of younger and older healthy adults to investigate whether object-location memory functions and maze learning differentially deteriorate with age, and whether this is due to age-related decline in working or recognition memory or spatial planning.

METHODS

Participants

Twenty-five young adults (12 males) between 20 and 30 were recruited, as well as 25 older adults between 60 and 70 (13 males). Education level was assessed using seven categories, 1 being the lowest (less than primary school) and 7 the highest (academic degree). Intelligence level was estimated using the Dutch version of the National Adult Reading Task (NART; Schmand, Lindeboom & Van Harskamp, 1992). Handedness was assessed with a modified version of the Annett Handedness Inventory (Briggs & Nebes, 1975), with scores between –24 (totally left handed) and +24 (totally right handed). None of the participants had a history of psychiatric or neurological disease (self report) and all older adults had a score above 27 on the Mini-Mental State Examination (MMSE; Folstein, Folstein & McHugh, 1975). Table 1 shows the background variables for both groups. Both age groups did not differ with respect to education level ($F(1,48) = 2.0$), sex distribution (Mann-Whitney $U = 247.5$, $Z = -1.5$) and handedness ($F(1,48) = 0.6$). The older group had a slightly higher estimated verbal IQ ($F(1,48) = 7.2$, $p < 0.01$), but absolute differences were small, with an estimated mean that was clearly within the 'average' intelligence range for both groups.

Materials

Virtual morris water maze

The vMWM (Behrens, 2007) is a computerised adaptation of the original Morris Water Maze used for rodent research (see Vorhees & Williams, 2006, for a detailed description). A blue circle is shown on a computer screen, indicating a birds-eye view of a swimming pool. Outside the circle, four objects are placed for orientation purposes, in a fixed position (see Figure 1). At the start of a trial, the mouse cursor is shown at a different location within the swimming pool. The participant subsequently has to move this cursor across the swimming pool in order to find a hidden platform, which is located at a fixed position within the pool. This platform is found if the cursor is close to its location and is made visible, after which the next trial is presented. The task consists of different conditions, i.e., the platform is rotated over subsequent trials, with the objects outside the pool and the platform remaining at the same relative location. Each task condition consists of three trials in which the maze is shown in one rotation angle, but with different starting points, presented in a fixed order to control for differential order effects (Allen & Larner, 1987). In the first condition, the maze is shown as in Figure 1 (0º rotation), in subsequent conditions, the maze is rotated 60º, 210º, 130º and 290º (resulting in a total of 15 trials). The participant was not told that the platform has a fixed location. To account for possible sensorimotor differences between participants due to differences in computer experience, the vMWM begins with a practice trial is shown in which the platform is visible (but at a different location than in the actual trials), and the participant is instructed to move the cursor to the platform using the shortest route possible. The computer registers the length of the performed routes in pixels per trial. The performance measure per trial is a ratio score, computed by dividing the actual route length by the shortest route length per trial (note that the shortest possible route length differs per trial due to the different starting point).

Object-location memory

The Object Relocation program (Kessels, Postma & De Haan, 1999) was used to present the different object location memory stimuli, in which ten everyday objects (e.g. ball, frog) were shown within a frame of 19 by 19 cm on the computer-screen. After a presentation time of 30 seconds, the objects disappeared from the display and were placed above the frame. Participants could locate them by clicking the objects and placing them within the frame using the mouse. The task consisted of three object-location memory conditions, each consisting of two different trials with new objects and new locations. First, in the *combined* condition, ten different objects were shown at different locations, which had to be relocated within an empty frame. This condition measures the ability to remember the conjunction of object identities and coordinate locations (see Figure 2A). In the *positional memory* condition, ten identical objects were shown at different locations in the presentation phase. Subsequently, these objects were presented above an empty frame and the participants had to relocate the objects as accurately as possible (see Figure 2B). This condition measures coordinate spatial memory. In the *object-location binding* condition, again ten different objects were shown at different locations, but in the relocation phase, the actual locations were marked by black dots. The participant had to assign the correct objects to these pre-marked locations (see Figure 2C). This condition relies on the ability to bind object identities to categorical positions (Kessels et al., 2002; Van Asselen et al., 2008).

For the combined condition, the absolute error score was computed in mm per trial, that is, the total distance between the objects' original positions and the relocated positions. For the object-location binding condition, the percentage incorrectly relocated objects was computed. For the positional memory condition, a best fit score was computed in mm, based on the paring of original to relocated positions that yielded the smallest distance error for the stimulus as a whole. This measure reflects the degree of positional reconstruction (see Kessels et al., 1999, for a more detailed discussion of these error measures).

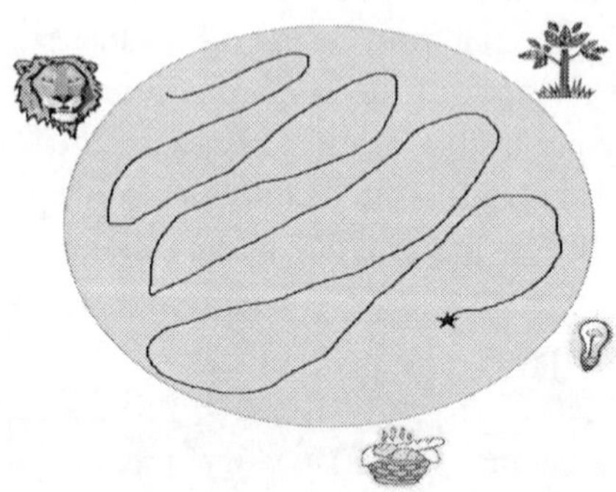

Figure 1. The virtual Morris Water Maze shown in the 0° angle with an example of a search path. The four objects outside the pool are shown for orientation purpose, the platform is indicated by the star (note that the platform was not visible for the participant until he or she was close to its location) .

Table 1. Mean age, education level, sex distribution, handedness and MMSE and NART-IQ for the young and older adults

	Younger adults (N = 25)		Older adults (N = 25)	
	Mean	SD	Mean	SD
Age	25.0	2.6	64.3	2.8
Education level	6.3	0.6	6.1	0.6
Handedness	18.0	9.9	20.2	9.5
Sex (m:f)	12:13		13:12	
MMSE			29.4	0.8
NART-IQ	103.3	5.6	109.3	9.7

Neuropsychological assessment

Standard neuropsychological tests were administered for the assessment of spatial working memory, visual recognition memory and spatial planning. The subtest Spatial Span of the Wechsler Memory Scale – 3rd edition (WMS-III) is an analogue of the Corsi Block-Tapping Task and measures visuospatial working memory (Wechsler, 1997). The task consist of 10 blocks that are fixed on a board. The experimenter taps block sequences of increasing length that have to be tapped by the participant in the same order (forward condition) or in reverse order (backward condition). The eventual score is the number of correctly tapped trials.

The Doors Test is a visual recognition-memory test that is part of the Doors and People Test (Baddeley, Emslie, & Nimmo-Smith, 1994). It consists of photographs of doors that have to be remembered and subsequently recognized among three distracter items. In the first subtest (A), 12 photographs of doors are presented serially at a rate of 1 per 3 seconds per photograph (using a booklet). Subsequently, 12 recognition items are shown, that each consist of four photographs of doors, among which one door that was previously presented. The three distractor doors are clearly distinct from the original door. The participant has to indicate the previously shown door. In subtest B, 12 new doors are presented, again followed by a recognition phase with new distracter items, but now the distracter items are visually highly similar to the to be remembered door. Each condition is preceded by a practice trial. The memory score is the number of correctly recognized doors for each subtest.

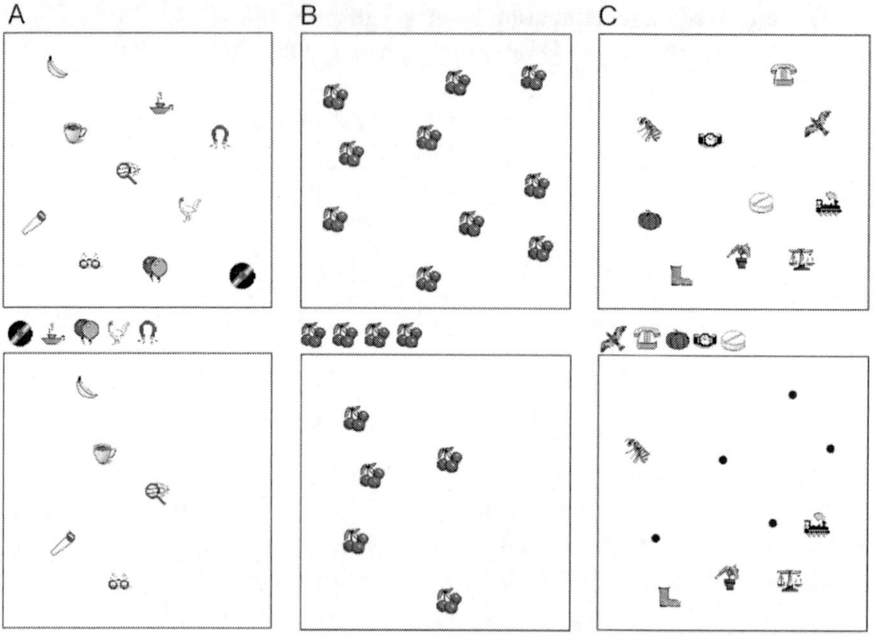

Figure 2. The three task conditions of the Object Relocation paradigm: A) combined, B) positional memory, and C) object-location binding.

The Key Search Test is a subtest of the Behavioural Assessment of the Dysexecutive Syndrome (BADS) battery (Wilson, Alderman, Burgess, Emslie & Evans, 1996). This test is an adaptation of a task originally developed by Terman and Merrill (1937) as part of the Stanford-Binet Intelligence Test. It consist of a 9.6 × 10 cm printed square, with a dot placed below the square to indicate the starting point. The participant is instructed to imagine that the square indicates the map of a field, that has to be searched in a way that the whole field is covered, beginning at the dot. The participant indicates the search path using a pencil. The test is scored using criteria described in detail in the test's manual (Wilson et al., 1996), with a score ranging from 0 to 16 (a higher score indicating a better performance).

Analyses

For the vMWM, a repeated-measure GLM analysis was performed with Trial (3 levels) and Angle (5 levels) as within-subject factor and Group (young vs. older) as between-subject factor. Between-group differences on the object-location memory conditions and the standard neuropsychological tests were analyzed using ANOVA. Effect sizes (Cohen's *d*) were computed for standardizing the between-group differences across the various tasks. Pearson's product-moment correlations coefficients (*r*) were computed between compound scores (standardized *z* scores) for combined object-location memory, positional memory, object-location memory, visual recognition memory (parts A and B taken together), spatial working memory (Spatial Span forward and backward taken together), maze learning (all trials and rotation angles taken together) and spatial planning that were computed for data-reduction purposes.

RESULTS

Virtual Morris Water Maze

Figure 3 shows the results for the different rotation conditions on the vMWM for the young and older adults. The repeated measures GLM revealed main effects of Trial ($F(2,47) = 11.5, p < 0.0005$) and Angle ($F(4,45) = 11.9, p < 0.0005$). No main effect of Group was found ($F(1,48) = 1.7$; overall effect size Cohen's $d = -0.43$). There were no significant interactions between Trial × Group ($F(2,47) = 2.0$), Angle × Group ($F(4,45) = 0.74$), Angle × Trial ($F(8,41) = 1.2$) or Angle × Trial × Group ($F(8,41) = 0.9$).

Object-location memory

Table 2 shows the results for the object-location memory task. ANOVA showed significant group differences on all task conditions: older participants had more incorrect relocations in the object-location binding condition ($F(1,48) = 10.9, p < 0.002, d = -0.93$), larger deviations on the combined condition ($F(1,48) = 9.8, p < 0.003, d = -0.86$) and remembered the overall spatial configuration, regardless of the objects, less accurately in the positional memory condition ($F(1,48) = 17.0, p < 0.0005, d = -1.17$).

Neuropsychological assessment

Table 2 also shows the performance for both groups on the standard neuropsychological tests. ANOVA showed that the older group performed worse than the young group on the Doors Test parts A ($F(1,48) = 6.8, p <$

0.02) and B ($F(1,48) = 15.6$, $p < 0.0005$), with an overall d-value of 1.16. Performance on the Spatial Span subtest of the WMS-III was worse for the older adults compared to the young group, both for the forward ($F(1,48) = 17.5$, $p < 0.0005$) and the backward ($F(1,48) = 6.4$, $p < 0.02$) condition (overall d-value = 1.40). No group differences were found on the BADS Key Search test ($F(1,48) = 2.8$, $d = 0.47$).

Table 3 shows the correlations (Pearson's r) between the scores on the different tasks, showing that the spatial-working memory test is highly correlated with all other tasks except spatial planning, that visual recognition memory is highly correlated with the object-location memory conditions, and that maze learning correlates with the object-location memory conditions that require fine-grained positional memory. Spatial planning was not correlated with any of the other task conditions.

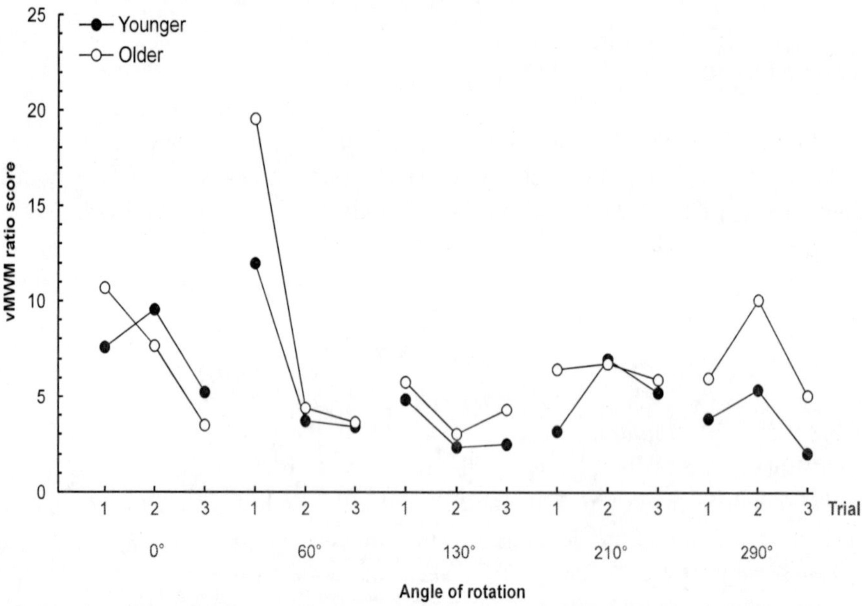

Figure 3. Performance for the young and older adults on the three trials for the different rotation conditions of the virtual Morris Water Maze (vMWM). The ratio score is the ratio between the actual path and the shortest possible path, with higher scores reflecting a worse performance.

Table 2. Mean performance (+ SD) on the standard neuropsychological tasks and the object-location memory conditions for the young and older adults

	Younger adults		Older adults	
	Mean	SD	Mean	SD
Doors Test				
Part A*	11.2	1.2	10.1	1.7
Part B***	9.0	1.8	7.0	1.9
Spatial Span (WMS-III)				
Forward***	8.6	1.8	6.5	1.8
Backward*	8.3	2.2	7.0	1.5
Key Search (BADS)	14.4	2.3	13.2	2.9
Object-location memory#				
Combined condition*	260.9	119.3	366.4	118.9
Object-location binding**	7.8	9.6	22.2	19.6
Positional memory***	166.8	39.9	221.4	52.9

* $p < 0.05$; ** $p < 0.005$; *** $p < 0.0005$; # higher scores indicate a worse performance

CONCLUSION

It was the aim of the present study to investigate possible differential effects of aging on spatial search (i.e., maze learning) and object-location memory, and the contribution of spatial working memory, visual recognition memory and spatial planning to spatial search and object-location memory. The results clearly show age-associated decline on object-location memory, visual recognition memory and spatial working memory, with effect sizes that can be classified as 'large' (Cohen, 1988). No differences between young and older adults were found on the virtual maze learning test and the spatial planning test.

With respect to memory for object locations, age effects are shown in all three task conditions. This is in agreement with previous studies also showing age-associated memory impairment on memory for object locations (Kessels et al., 2007) or positional memory (Siedlecki, 2007). The question is whether this object-location memory decline is specific, or whether it reflects a general age-associated memory decline (see, for example, Bastin and Van der Linden, 2006). Our findings also show age differences on a task measuring visual recognition memory, i.e. the Doors Test, and high correlations between the object-location memory conditions and visual working memory and recognition memory tasks were found. Although previous studies have indicated that object-location memory declines to a greater degree than visual memory for items, the present study shows age effects on these tasks that have similar effect sizes. Alternatively, it can be argued that our object-location memory tests, which present static information, are predominantly visual in nature. That is, some authors state that spatial memory tasks should contain dynamic aspects, such as a temporal order of locations or moving objects (Zimmer, Speiser & Seidler, 2003; Logie, 1995). However, while the present spatial memory conditions used static information in the presentation phase, the relocation phase of the task conditions did require active spatial manipulation of the individual objects. The spatial memory conditions thus clearly differ in this regard from the recognition task used to assess visual recognition memory. In sum, the present results showing similar effect sizes between young and older adults and strong correlations between object-location memory, working memory and recognition memory indicate that the age-related decline in object-location memory is in large part due to an underlying decline in overall memory function and not limited to spatial-memory deterioration.

Interestingly, we did not observe aging effects on the vMWM. Only in two previous studies was age-related decline on a two-dimensional vMWM examined (Driscoll et al., 2004; Moffat & Resnick, 200), demonstrating significant age-related decline. Apart from the task-related differences already outlined in the Introduction, other methodological differences exist that may explain this discrepancy. First, in the Driscoll et al. (2007) study, young adults were university students and the older adults (>60) were community-dwelling healthy older people not affiliated with the university.

Although no information is provided with respect to educational background or intelligence estimates, it is unlikely that these two groups were matched on education level, as was done in the present study. Consequently, the difference between the young and older adults may also have been due to differences in intelligence which is generally one of the important confounding variable when examining cognitive function (Lezak, Howieson & Loring, 2004). Second, in the study by Moffat and Resnick (2002), older participants were between 60 and 93, whereas in our study, they were all between 60 and 70. Indeed, Moffat and Resnick (2002) found that up to 65 years of age, performance differences between older and young adults on the vMWM were small, but they reported a sharp nonlinear performance decline after the age of 65.

Table 3. Correlations (Pearson's r) between the different tasks and task conditions

	Object-location binding	Combined object-location memory	Positional memory	Visual recognition memory	Spatial working memory	Maze learning
Combined object-location memory	0.63**					
Positional memory	0.49**	0.43**				
Visual recognition memory	-0.47**	-0.43**	-0.42**			
Spatial working memory	-0.47**	-0.54**	-0.52**	0.42*		
Maze learning	0.27	0.33*	0.46**	-0.27	-0.37*	
Spatial planning	-0.08	0.01	-0.08	-0.03	0.20	0.15

* $p < 0.05$; ** $p < 0.01$

Note. Negative correlations are the result of reversed scales.

Hence, it may well be that in our sample, although age-associated decline was consistently found on object-location memory, spatial working memory and visual recognition memory, spatial search remains intact until the later decades of the life span (i.e., >70). Future studies should examine this possibility in more detail. Alternatively, one could argue that this paradigm was too easy for both young and older adults, since the maze learning paradigm uses only one fixed location. However, clear learning curves were found after each rotation for both groups, making this a less plausible explanation.

We also did not observe age differences on the Key Search test, part of an executive function test battery, requiring spatial planning (Wilson et al., 1996). There is abundant evidence that these aspects of executive function deteriorate as a function of age, as measured with tasks requiring planning of routes in a maze or on a map (Allain et al., 2005; Salthouse & Siedlecki, 2006). However, although these tests heavily rely on complex, rule-based planning, they do not require the ability to systematically scan and search an environment, a process we refer to here as spatial planning. The Key Search test, in contrast, primarily addresses the latter process. To our knowledge, ours is the first study to examine possible age effects on the Key Search test, showing that older adults are clearly not impaired in their ability to plan a spatial scanning and search strategy. This may also explain the intact spatial search performance on the vMWM.

The largest age differences were found on the spatial working-memory task. Previous studies already demonstrated consistent age effects on spatial span measures (Wilde, Strauss & Tulsky, 2004; Robbins, James, Owen, Sahakian, Lawrence, McInnes & Rabbit, 1998) and large age effects have also been reported on spatial working-memory tasks that do not use the span format (Cornoldi et al., 2006; Robbins et al., 1998). It should be mentioned that all spatial-working memory paradigms typically require the integration of spatially or serially presented visual information, and thus may also rely on processes that are important for the binding of different features within working memory, such as mediated by the episodic buffer (Baddeley, 2007). Indeed, in our spatial span task, the binding of temporal order and spatial location is required, which may also be differentially affected by age. Still, the current findings are in agreement with the notion that working-memory decline is one of the most prominent characteristics of cognitive aging. This decline can be linked to age-related prefrontal cortical changes (Kessels, Postma, Wijnalda & De Haan, 2000; Tisserand & Jolles, 2003), resulting in a heightened susceptibility for distraction and a performance in older adults that is affected by memory load to a greater degree than in young adults (Gazzaley,

Sheridan, Cooney & D'Esposito, 2007). Furthermore, spatial working-memory performance was highly correlated with both the object-location memory tasks and maze learning. This indicates that a suboptimal spatial working-memory functioning at older age directly affects performance on other spatial memory and learning tasks.

While our findings clearly show that spatial working memory deteriorates substantially in comparison to other aspects of spatial cognition, alternative factors may also play a role. For example, there are methodological differences between the various tasks that may be relevant. That is, in the object-location memory task conditions ten objects or ten locations have to be remembered. This is in contrast to the maze-learning task, in which only one location has to be remembered. Furthermore, although spatial working memory may be regarded as a prerequisite for both tasks, the object-location memory tasks predominantly tap the more 'passive' visuospatial sketchpad, whereas maze learning may also rely on the active central-executive processes within working memory (Baddeley & Hitch, 1974). However, although our results indeed show a correlation between spatial working memory and the vMWM, we did not observe age-related decline on the latter task, suggesting that the vMWM may have been not demanding enough to recruit central-executive working-memory processes to a large extent. Also, the number of measures or trials across tasks differs. The Key Search test, for example, consists of only one single trial, whereas all other tasks use multiple trials or measures. Although multiple trials may generally render the performance on a tasks more reliable, assessment of executive function – where the novelty of a task is crucial – is generally hampered by repeated testing. Furthermore, it is unlikely that differences in complexity can account for all findings, such as the larger effect size for the spatial span task compared to the spatial memory tasks. Alternatively, it may be argued that age-associated decline on memory tasks is the result of diminished speed of information processing (Salthouse, 1996) and not due to a memory impairment in itself. Indeed, numerous studies have shown that information-processing speed declines as a function of age (see, e.g. Finkel, Reynolds, McArdle & Pedersen, 2007). However, for the present study we deliberately selected tasks that are not determined by a lowered information-processing speed. That is, none of our tasks applied restrictions in time for responding, and the presentation duration for the object-location memory tasks was sufficiently long (i.e., 30 seconds). The processing-speed hypothesis of aging may partly explain the absence of age differences on the vMWM compared to previous studies showing age effects on this task though. That is, in the study by Moffat and Resnick (2002) older people were indeed slower than young adults in reaching the platform, but also traveled longer distances. Although time was not a performance measure in the study by Driscoll et al. (2005), their vMWM consisted of a search path that was set in motion by the participant, who had subsequently only control of the search path's direction, but not its speed. Consequently, age differences in both studies may have been the result of differences in processing speed. Since in our vMWM timing was not critical, this may explain the differential results.

In all, this is the first study to examine age effects on various aspects of spatial search and spatial memory in combination. Our results show large age effects on memory for object locations (including object-location binding and memory for positional information), spatial working memory and visual recognition memory. No age differences were found on spatial search as assessed by the vMWM and a non-memory spatial planning task. Thus, while spatial search and spatial planning are unimpaired in older adults, the age-associated decline on the spatial span and object-location memory tasks is in accordance with the notion that older people have more difficulties with increasing mental load and may be more susceptible to distraction or interference (i.e., the working-memory hypothesis of aging), possibly due to prefrontal dysfunction. It would be interesting for future studies to contrast these results with the performance of patients with hippocampal dysfunction, such as in amnestic Mild Cognitive Impairment (MCI) or early Alzheimer's disease. Also, the vMWM performance in older healthy adults (i.e., >70) may deteriorate to a greater extent compared to our sample. Finally, the 'birds-eye' viewpoint used in most human analogues of the vMWM cannot be used to study egocentric processing. Using a three-dimensional virtual-reality environment in addition to the standard allocentric task may be an interesting extension of the present set-up to study age-related decline on maze learning.

REFERENCES

Allain, P., Nicoleau, S., Pinon, K., Etcharry-Bouyx, F., Barré, J., Berrut, G., Dubas, F. & Le Gall, D. (2005). Executive functioning in normal aging: A study of action planning using the Zoo Map Test. *Brain and Cognition*, *57*, 4-7.

Allen, E. A. & Larner, S. (1987). Spatial memory and perception in stroke patients: The importance of order effects. *British Journal of Clinical Psychology, 26*, 227-8

Baddeley, A. D., Emslie, H. & Nimmo-Smith, I. (1994). *Doors and People: A test of visual and verbal recall and recognition*. Bury St. Edmunds, UK: Thames Valley Test Company.

Baddeley, A. D. & Hitch, G. J. (1974). Working memory. In G.A. Bower (Ed.), *Recent advances in learning and motivation*, Vol. *8*, (47-89). New York: Academic Press.

Baddeley, A. (2007). *Working memory, thought, and action*. Oxford: Oxford University Press.

Bastin, C. & Van der Linden, M. (2006). The effects of aging on the recognition of different types of associations. *Experimental Aging Research, 32*, 61-77.

Begega, A., Cienfuegos, S., Rubio, S., Santin, J. L., Miranda, R. & Arias, J. L. (2001). Effects of ageing on allocentric and egocentric spatial strategies in the Wistar rat. *Behavioural Processes*, *53*, 75-85.

Behrens, M. (2007). *Der Morris Water Maze Test: Entwicklung und Anwendung eines neuen Verfahrens zur Differenzialdiagnostik organisch bedingter Gedächtnisstörungen* [The Morris Water Maze test: Development and application of a new test for the differential diagnosis of organic memory disorders]. Unpublished Master thesis, University of Oldenburg, Germany.

Briggs, G. G. & Nebes, R. D. (1975). Patterns of hand preference in a student population. *Cortex, 11*, 230-238.

Brown, H. D. & Kosslyn, S. M. (1993). Cerebral lateralization. *Current Opinion in Neurobiology, 3*, 183-186.

Chalfonte, B. L., Verfaellie, M., Johnson, M. K. & Reiss, L. (1996). Spatial location memory in amnesia: Binding item and location information under incidental and intentional encoding conditions. *Memory, 4*, 591-614.

Cherry, K. E., Park, D. C. & Donaldson, H. (1993). Adult age differences in spatial memory: Effects of structural context and practice. *Experimental Aging Research, 19*, 333-350.

Clark, R. E., Broadbent, N. J. & Squire, L. R. (2007). The hippocampus and spatial memory: Findings with a novel modification of the water maze. *Journal of Neuroscience, 27*, 6647-6654.

Cohen, J. (1988). *Statistical power analysis for the behavioral sciences* (2nd ed.). Hillsdale, NJ: Erlbaum.

Cooney, R. & Arbuckle, T. (1997). Age, context, and spatial memory: A neuropsychological approach. *Aging, Neuropsychology, and Cognition, 4*, 249-265.

Cornoldi, C., Bassani, C., Berto, R. & Mammarella, N. (2007). Aging and the intrusion superiority effect in visuo-spatial working memory. *Aging, Neuropsychology, and Cognition, 14*, 1-21.

D'Hooge, R. & De Deyn, P. P. (2001). Applications of the Morris water maze in the study of learning and memory. *Brain Research Reviews, 36*, 60-90.

Driscoll, I., Hamilton, D. A., Yeo, R. A., Brooks, W. M. & Sutherland, R. J. (2005). Virtual navigation in humans: the impact of age, sex, and hormones on place learning. *Hormones and Behavior, 47*, 326-335.

Feigenbaum, J. D. & Morris, R. G. (2004). Allocentric versus egocentric spatial memory after unilateral temporal lobectomy in humans. *Neuropsychology, 18*, 462-472.

Finkel, D., Reynolds, C. A., McArdle, J. J. & Pedersen, N. L. (2007). Age changes in processing speed as a leading indicator of cognitive aging. *Psychology and Aging, 22*, 558-568.

Fitting, S., Allen, G. L. & Wedell, D. H. (2007). Remembering places in space: A human analogue study of the Morris Water Maze. In: T., Barkowsky, M., Knauff, G. Ligozat, & D. R. Montello (Eds.), *Spatial cognition V: Reasoning, action, interaction* (59-75). Berlin: Springer.

Fjell, A. M., Walhovd, K. B., Reinvang, I., Lundervold, A., Dale, A. M., Quinn, B. T., Makris, N. & Fischl, B. (2005). Age does not increase rate of forgetting over weeks: Neuroanatomical volumes and visual memory across the adult life-span. *Journal of the International Neuropsychological Society, 11*, 2-15.

Folstein, M. F., Folstein, S. E. & McHugh, P. R. (1975). "Mini-mental state": A practical method for grading the cognitive state of patients for the clinician. *Journal of Psychiatric Research, 12*, 189-198.

Gazzaley, A., Sheridan, M. A., Cooney, J. W. & D'Esposito, M. (2007). Age-related deficits in component processes of working memory. *Neuropsychology, 21*, 532-539.

Karnath, H. O. & Wallesch, C. W. (1992). Inflexibility of mental planning: A characteristic disorder with prefrontal lobe lesions? *Neuropsychologia, 30,* 1011-1016.

Kessels, R. P. C., Hendriks, M. P. H., Schouten, J., Van Asselen, M. & Postma, A. (2004). Spatial memory deficits in patients after unilateral amygdalohippocampectomy. *Journal of the International Neuropsychological Society, 10*, 907-912.

Kessels, R. P. C., Hobbel, D. & Postma, A. (2007). Aging, context memory and binding: A comparison of "what, where and when" in young and older adults. *International Journal of Neuroscience, 117*, 795-810.

Kessels, R. P. C., Kappelle, L. J., De Haan, E. H. F. & Postma, A. (2002). Lateralization of spatial-memory processes in humans: Evidence on spatial span, maze learning and memory for object locations. *Neuropsychologia, 40*, 1465-1473.

Kessels, R. P. C., Postma, A. & De Haan, E. H. F. (1999). Object Relocation: A program for setting up, running, and analyzing experiments on memory for object locations. *Behavior Research Methods, Instruments, & Computers, 31*, 423-428.

Kessels, R. P. C., Postma, A., Wijnalda, E. M. & De Haan, E. H. F. (2000). Frontal-lobe involvement in spatial memory: Evidence from PET, fMRI, and lesion studies. *Neuropsychology Review, 10*, 101-113.

Kessels, R. P. C., Te Boekhorst, S. & Postma, A. (2005). The contribution of implicit and explicit memory to the effects of errorless learning: A comparison between young and older adults. *Journal of the International Neuropsychological Society, 11*, 144-151.

Lezak, M. D., Howieson, D. B. & Loring, D. W. (2004). *Neuropsychological assessment* (4th ed). Oxford, Oxford University Press.

Logie, R. H. (1995). *Visuo-spatial working memory*. Hove, UK: Lawrence Erlbaum Associates.

Moffat, S. D. & Resnick, S. M. (2002). Effects of age on virtual environment place navigation and allocentric cognitive mapping. *Behavioral Neuroscience, 116*, 851-859.

Mou, W., McNamara, T. P., Rump, B. & Xiao, C. (2006). Roles of egocentric and allocentric spatial representations in locomotion and reorientation. *Journal of Experimental Psychology: Learning, Memory, and Cognition, 32*, 1274-1290.

Murphy, G. G., Rahnama, N. P. & Silva, A. J. (2006). Investigation of age-related cognitive decline using mice as a model system: Behavioral correlates. *American Journal of Geriatric Psychiatry, 14*, 1004-1011.

O'Keefe, J. & Nadel, L. (1978). *The hippocampus as a cognitive map*. Oxford, UK: Oxford University Press.

Perlmutter, M., Metzger, R., Nezworski, T. & Miller, K. (1981). Spatial and temporal memory in 20 to 60 year olds. *Journal of Gerontology, 36*, 59-65.

Piekema, C., Fernández, G., Postma, A., Hendriks, M. P. H., Wester, A. J. & Kessels, R. P. C. (2007). Spatial and non-spatial contextual working memory in patients with diencephalic or hippocampal dysfunction. *Brain Research, 1172*, 103-109.

Piekema, C., Kessels, R. P. C., Mars, R. B., Petersson, K. M. & Fernández, G. (2006). The right hippocampus participates in short-term memory maintenance of object-location associations. *Neuroimage, 33,* 374-382.

Postma, A., Kessels, R. P. C. & Van Asselen, M. (2004). The neuropsychology of object-location memory. In G. L. Allen, (Eds.), *Human spatial memory: Remembering where* (pp. 143-160). Mahweh, NJ: Lawrence Erlbaum Associates.

Postma, A., Kessels, R. P. C. & Van Asselen, M. (2008). How the brain remembers and forgets where things are: The neurocognition of object-location memory. *Neuroscience and Biobehavioral Reviews, 32*, 1339-1345.

Robbins, T. W., James, M., Owen, A. M., Sahakian, B. J., Lawrence, A. D., McInnes, L. & Rabbitt, P. M. (1998). A study of performance on tests from the CANTAB battery sensitive to frontal lobe dysfunction in a large sample of normal volunteers: Implications for theories of executive functioning and cognitive aging. *Journal of the International Neuropsychological Society, 4*, 474-490.

Rogers, J. L. & Kesner, R. P. (2006). Lesions of the dorsal hippocampus or parietal cortex differentially affect spatial information processing. *Behavioral Neuroscience, 120*, 852-860.

Salthouse, T. A. (1996). The processing-speed theory of adult age differences in cognition. *Psychological Review, 103*, 403-428.

Salthouse, T. A. & Siedlecki, K. L. (2007). Efficiency of route selection as a function of adult age. *Brain and Cognition, 63*, 279-286.

Schmand, B., Lindeboom, J. & Van Harskamp, F. (1992). *Nederlandse Leestest voor Volwassenen* [Dutch Adult Reading Test]. Lisse, The Netherlands: Swets & Zeitlinger.

Seo, E. H., Lee, D. Y., Choo, I. H., Youn, J. C., Kim, K. W., Jhoo, J. H., Suh, K. W., Paek, Y. S., Jun, Y. H. & Woo, J. I. (2007). Performance on the Benton Visual Retention Test in an educationally diverse elderly population. *Journals of Gerontology, 62B*, P191-P193.

Siedlecki, K. (2007). Investigating the structure and age invariance of episodic memory across the adult lifespan. *Psychology and Aging, 22*, 251-268.

Terman, L. M. & Merrill, M. A. (1937). *Measuring Intelligence*. Boston: Houghton Mifflin.

Tisserand, D. J. & Jolles, J. (2003). On the involvement of prefrontal networks in cognitive ageing. *Cortex, 39*, 1107-1128.

Van Asselen, M., Kessels, R. P. C., Kappelle, L. J. & Postma, A. (2008). Categorical and coordinate spatial representations within object-location memory. *Cortex, 44*, 249-256.

Vorhees, C. V. & Williams, M. T. (2006). Morris water maze: Procedures for assessing spatial and related forms of learning and memory. *Nature Protocols, 1*, 848-858.

Wager, T. D. & Smith, E. E. (2003). Neuroimaging studies of working memory: A meta-analysis. *Cogntive, Affective, and Behavioral Neuroscience, 3*, 255-274.

Wechsler, D. (1997a). *Wechsler Memory Scale – Third Edition: Manual*. San Antonio, TX: Psychological Corporation.

Weniger, G., Ruhleder, M., Wolf, S., Lange, C. & Irle, E. (2009). Egocentric memory impaired and allocentric memory intact as assessed by virtual reality in subjects with unilateral parietal cortex lesions. *Neuropsychologia, 47*, 59-69.

Wilde, N., Strauss, E. & Tulsky, D. S. (2004). Memory span on the Wechsler scales. *Journal of Clinical and Experimental Neuropsychology, 26*, 539-549.

Wilson, B. A., Alderman, N., Burgess, P., Emslie, H. & Evans, J. (1996). *Behavioural Assessment of the Dysexecutive Syndrome (BADS)*. Bury St. Edmonds, UK: Thames Valley Test Company.

Yamada, K., Kawai, H., Iwasaki, T. & Ichitani, Y. (2002). Residential maze as a task for testing rats' maze learning ability: Effects of hippocampal lesions and cholinergic receptor antagonists. *Neuroreport, 13*, 1759-1764.

Yoon, T., Okada, J., Jung, M. W. & Kim, J. J. (2008). Prefrontal cortex and hippocampus subserve different components of working memory in rats. *Learning and Memory, 15*, 97-105.

Zaehle, T., Jordan, K., Wustenberg, T., Baudewig, J., Dechent, P. & Mast, F. W. (2007). The neural basis of the egocentric and allocentric spatial frame of reference. *Brain Research, 1137*, 92-103.

Zimmer, H. D., Speiser, H. R. & Seidler, B. (2003). Spatio-temporal working-memory and short-term object-location tasks use different memory mechanisms. *Acta Psychologica, 114*, 41-65.

In: Spatial Memory ISBN: 978-1-61668-139-5
Editor: Jacob B. Thomas, pp. 101-121

Chapter 4

THE ROLE OF STEROID HORMONES IN BRAIN DEVELOPMENT AND VISUOSPATIAL PROCESSING ON ADOLESCENCE

Cristian Gurzu
National College Nicolae Balcescu, Braila, Romania
University "Al. I. Cuza" of Iasi, Romania
Department of Molecular and Experimental Biology, Biology Faculty

ABSTRACT

One of the central aims in developmental cognitive neuroscience, is to relate changes in cognition to maturational changes in the brain. Research with animals has been helpful in understanding links between brain changes during development as they relate to cognition. A good example comes from studies of spatial memory in relation to the development of the hippocampus. Before 19 days of age, rats pups are unable to solve the hidden platform version of the Morris watter maze task. By 21 days of age, however, rats can typically solve the Morris water maze (Purves et al., 2008). In this structure were found a high density of steroid receptors, suggesting there must be some relationship between steroid receptors and development of hippocampus (Simerly et al., 1990).

Adolescence is now recognized as a major and dynamic period of neural development during which behavioral circuits are remodeled and refined. Although the human brain of a 5-year old child is already 90% of its adult size (Dekaban, 1978), significant remodeling is still to come. This concept was kindled by research in both humans and animals documenting that many of the basic developmental processes occurring during perinatal brain development are recapitulated during adolescence. Compared with activation of neural circuits by steroid hormones, it is relatively little known about steroid-dependent organization during adolescence. Presumably, there are many similarities to the processes involved in steroid-dependent sexual differentiation during early neural development, such as regulation of cell death and survival, and synaptic density and connectivity.

During the adolescent phase of organization, steroid-dependent refinement of neural circuits results in long-lasting structural changes in brain, which determine adult behavioral responses to hormones or sensory stimuli and improve the cognitive abilities such as attention, learning and memory, reasoning and visuospatial processing, as adaptative response to environment and social interaction.

The organizational changes in the adolescent brain under steroid hormones could evolve in a progressive manner, in order to respond from simple to complex stimuli and thus, the trajectory of cognitive development follow the developmental course of brain neuromaturation, from visual perception and spatial memory to visuospatial processing and semantic association between visual informations and abstract categories.

ADOLESCENCE AS A BRAIN SENSITIVE PERIOD FOR STEROID HORMONES

Adolescence is an important developmental period in which major physical, psychological, cognitive, and social transformations occur (Blakemore and Choudhury, 2006; Galvan et al., 2007; Markham et al., 2007) and gender differences emerge and manifest themselves (Lenroot et al., 2007; Schmithorst et al., 2007). Behavioral transformations are closely related to cerebral development, encompassing dramatic and widespread changes in brain morphology, under gonadal steroid hormones.

Although basic developmental processes are comparable between boys and girls, such as linear increases in white matter volume and nonlinear inverted u-shaped development of regional gray matter, sexual dimorphisms have been reported for global and regional brain volumes and the time course of brain development (Giedd et al., 1997, 1999).

Whereas peaks in gray matter volume typically occur 1 year earlier in girls than in boys, the rates of global volume changes follow a steeper slope in boys than in girls with respect to both increase of white matter and decrease in gray matter volumes.

Likewise, volumes of subcortical structures like the striatum, the hippocampus, and the amygdala change during adolescence in a sex-dependent manner, with amygdala volume increasing significantly more in males than in females and hippocampal and striatal volume increasing more in females (Caviness et al.,1996; Giedd et al.,1997; Lange et al., 1997). Although it has recently been demonstrated that sexual dimorphism in total cerebral and subcortical gray matter and white matter volumes are already present in the neonatal brain (Gilmore et al., 2007), many sex differences in brain structures seem to occur after the age of 9 or 10 (Goldstein et al., 1999; Giedd et al., 2006). Recently, De Bellis et al.(2001) demonstrated that brain maturation is not only affected by significant interaction effects between sex and age but also by interactions between sex and measures of physical maturation (De Bellis et al., 2001) as assessed by Tanner stages (TS, Tanner and Davies, 1985).

It is believed that the development of gray matter volume in particular more or less coincides with the onset of puberty (De Bellis et al., 2001; Sowell et al., 2002; Gogtay et al., 2004). The increase of gray matter volume at the onset of puberty might reflect a coincident wave of synaptic proliferation and the gradual decrease in gray matter density that takes place after puberty, might be attributed to postpubescent synaptic pruning (Bourgeois et al., 1994).

The remodeling of the adolescent brain is accomplished through a variety of mechanisms, including neurogenesis (Eckenhoff and Rakic, 1988; He and Crews, 2007; Pinos et al., 2001; Rankin et al., 2003), programmed cell death (Nunez et al., 2001, 2002), elaboration and pruning of dendritic arborizations and synapses (Andersen et al., 1997; Huttenlocher and Dabholkar, 1997; Lenroot and Giedd, 2006; Sowell et al., 2004), myelination (Benes et al., 1994; Paus et al., 2001; Sowell et al., 2001), and sexual differentiation (Chung et al., 2002; Davis et al., 1996; Nunez et al., 2001). These processes are known to be influenced by both androgens (MacLusky et al., 2006) and estrogens (Galea et al., 2006).

The effects of sex steroids on brain morphology are classically described as operating through two distinct mechanisms (Cooke et al., 1998). One mechanism, called „organization", is defined as a developmental mechanism in which steroids act during critical periods to mediate permanent sexually dimorphic differentiation of brain morphology that gives rise to male and female sexual behavior and physiology in adulthood. The other mechanism, called „activation", is mediated through the acute effects of gonadal hormones on the fully developed nervous system and is responsible for maintaining sex-specific behaviors in adulthood (Breedlove and Hampson, 2002).

Although the organization-activation framework for steroid control of reproductive behavior originally presumed a strictly activating role for gonadal steroids during adolescence, a recent modernization of this thinking incorporates dual roles for steroid hormones, proposing that they not only activate but also organize neural circuits during adolescence (Romeo et al., 2002).The sequence of events during steroid-dependent adolescent maturation of reproductive behavior may be an initial reorganization of circuits that further sensitizes them to hormone activation (Sisk and Foster, 2004).

Obviously, the adolescent brain maturation under steroid hormones involves the development of other cognitive abilities which improve the social interaction, communication, competition and complex behaviors than innate reproductive behavior.

Identifying the structural correlates of behavioral maturation and determining which structural features are related to steroid hormonal changes is an important area for further research.

Brain Development on Adolescence

During adolescence cortical gray matter volume tends to follow an „inverted U" developmental course with volumes peaking at different times in different lobes (Lenroot and Giedd, 2006). For instance, frontal lobe gray matter reaches its maximal volume at 11.0 years in girls and 12.1 years in boys; temporal lobe cortical gray matter peaks at 16.7 years in girls and 16.2 years in boys; and parietal lobe cortical gray matter peaks at 10.2 years in girls and 11.8 years in boys (Giedd et al. 1999b).

The developmental trajectory of cortical gray matter followed a regionally specific pattern with areas subserving primary functions, such as motor and sensory systems, maturing earliest and higher order association areas, which integrate those primary functions, maturing later. For example, in the temporal lobes the latest part to reach adult levels is the superior temporal gyrus/sulcus which integrates memory, audio-visual input, and object recognition functions, along with prefrontal and inferior parietal cortices (Mesulam, 1998; Calvert, 2001; Martin and Chao, 2001). Notably, late to reach adult levels of cortical thickness is the dorsolateral prefrontal cortex, involved in circuitry subserving control of impulses, judgment, decision making and visual processing

(Lenroot and Giedd, 2006; Nieder et al., 2002; Fuster 1989; Miller and Cohen, 2001).

In contrast to the inverted U shape of gray matter developmental curves, the amount of white matter in the brain generally increases throughout childhood and adolescence.

Unlike the lobar differences seen in gray matter trajectories, the white matter slopes are similar in frontal, temporal, and parietal lobes (Lenroot and Giedd, 2006).

The differences in gray matter and white matter developmental trajectories belie the inseparable connection among neurons, glial cells, and myelin, which are components of the same neural circuits and share lifelong reciprocal relationships (Fields and Stevens-Graham, 2002).

Neuron activity influences myelin production and the proliferation and survival of oligodendrocytes (Barres and Barde, 2000; Fields et al., 2001) while oligodendrocytes influence neurons via secretion of neuronal growth factors and influence axonal growth and clustering of ion channels (Du and Dreyfus, 2002). Proximal pathways tend to be myelinated before distal, sensory before motor, and projection before association (Volpe, 2000).

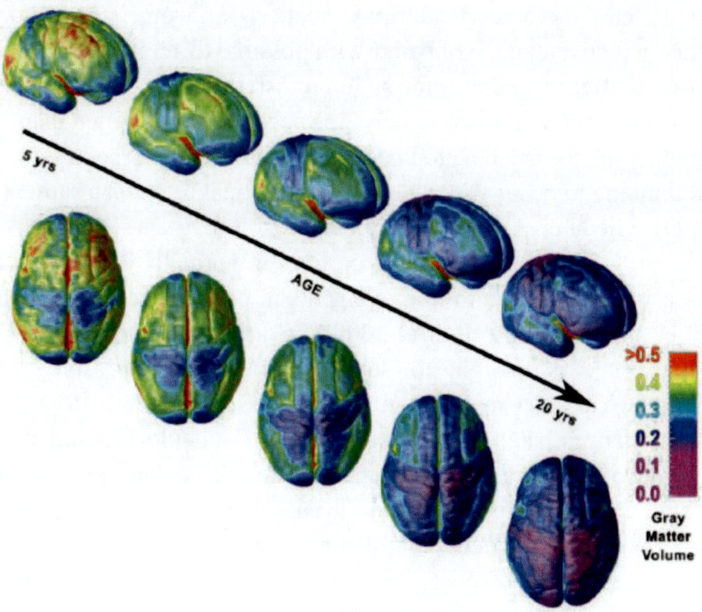

Right lateral and top views of the dynamic sequence of gray matter maturation over the cortical surface (Lenroot and Giedd, 2006).

Later-maturing myelin sheaths, such as those in association tracts and intracortical regions, tend to be thinner with greater axonal load per oligodendrocyte (Yakovlev and Lecours, 1967; Kinney et al., 1994), which may render them more vulnerable to environmental or aging-related factors (Bartzokis, 2004).

It shows that the cortical ribbon is actually thicker in some brain regions in females, despite the fact that females tend to have smaller bodies, and smaller brains, including smaller overall gray and white matter volumes than do men. The regions in which cortices are most prominently thicker in females than in males are in the right hemisphere association cortices, particularly the temporal and parietal lobes. In these regions, the cortex is up to 0.45 mm thicker in women; permutation analyses confirmed that these effects were not attributable to chance. The nonsignificant age-by-sex interactions in these regions suggest that the sex differences in temporoparietal cortices are stable across the life span (Sowell et al. 2007).

Regional cortical thickening in women may also be consistent with the profile of cognitive differences observed between the sexes, particularly the female advantage on language tasks that may be attributable to their thicker cortices in posterior perisylvian language regions. Gur and others (1999) have suggested that more cortical gray matter in women may provide a computational advantage, compared with possible differences in white matter, which would affect speed of information transfer.

A thicker cortex, however, may not necessarily be better than a thinner one. Studies of normal development, for example, have consistently shown cortical thinning to occur with age as part of normal brain maturation (Jernigan et al., 1991; Giedd et al., 1999; Sowell et al., 2004).

Thinning of frontal and parietal cortices in normally developing children, moreover, is associated with improvements in performance on language tasks (Sowell et al., 2001; Sowell et al., 2004). Cortical thinning during childhood and adolescence is thought to derive both from progressive changes in myelination (Yakovlev and Lecours, 1967; Benes et al., 1994) and from regressive changes, such as synaptic pruning (Huttenlocher and de Courten, 1987). Both of these cellular changes are thought to improve computational speed and efficiency as redundant synapses are eliminated and oft-used cortical circuits are insulated with myelin.

The parietal cortex subserves visuospatial functions, and thus, thinner cortices in men within temporoparietal regions, if it indeed arises from greater pruning and myelination during development and more efficient computational processing, may contribute to the superior visuospatial skills of men.

Differences in the rates of cortical maturation between males and females, and relationships between cortical maturation and changes in cognitive function, could be assessed across the cerebrum.

STEROID HORMONES AND DEVELOPING OF COGNITIVE ABILITIES

Studies have identified weak but significant differences between the overlapping distributions of male and female abilities, with women reported to perform better as a group than men on verbal, perceptual, and fine motor tasks, but poorer than men on spatial and quantitative measures (Sherwin, 1994).

As suggested by studies in other species, differences in brain structure established during development by the presence or absence of androgen probably account for performance differences between the sexes (Nass and Baker,1991; Naghdi et al., 2001; Gurzu et.al, 2008). Based on evidence from studies of adults, it has been suggested that androgen levels correlate negatively with performance on verbal tasks, but positively with performance on spatial tasks. Endogenous testosterone levels determined from plasma or saliva samples have been positively correlated with the performance of men on spatial measures (Christiansen and Knussmann, 1987; Christiansen,1993, Errico et al., 1992) and testosterone supplementation selectively improved spatial performance in elderly men and female-to-male transsexuals (Janovsky, 1994; van Goozen et al., 1995). There also is evidence that endogenous and supplemental testosterone impaired performance on some verbal measures (Christiansen and Knussmann, 1987; Christiansen,1993; Phillips and Sherwin, 1992a; van Goozen et al., 1995), but improved the object location (Postma et al., 2000).

Available data indicate that endogenous or exogenous estrogen enhances performance on tasks that depend primarily on working memory, but usually fails to alter or even impairs, performance of reference-memory tasks (Daniel et al., 1997; Chesler and Juraska, 2000).

Both androgens and estrogens support the maintenance of synapses in hippocampus and prefrontal cortex in animal models. For instance, gonadectomy or ovariectomy causes a profound loss of synapse density in hippocampus of male and female rats (Leranth et al.,2003; Wooley and McEwen, 1994) and monkeys (Leranth et al.,2004).

During adolescence and this time of active neural maturation, many cognitive processes are also developing. One such process is working memory. Working memory refers to the ability to actively store and manipulate information online over brief periods of time (Baddeley, 1986).This ability is fundamental to intact performance in a variety of other cognitive domains, including language comprehension, abstract reasoning, learning and memory and visuospatial processing (Baddeley, 1992; Gathercole, 1999; Vuontela, 2003).

Verbal and spatial working memory abilities improve throughout childhood and adolescence (Gathercole et al., 2004; Luna et al., 2004), with accuracy and reaction times increasing and decreasing respectively during spatial n-back (Kwon et al., 2002; Vuontela et al., 2003) and spatial delayed response tasks (Zald and Iacono, 1998). It is likely that these behavioral improvements in working memory are the result of the described neuromaturational processes that are occurring during the child and adolescent years.

In a recent morphometric study, the most pronounced gray matter loss was found in the parietal lobes in normally developing children for both sexes (Wilke et al., 2007). Thus, the negative correlations between parietal structures and testosterone in boys and girls might be considered within the context of the general decrease in parietal gray matter volumes during normal development.

The development of gray matter in the parietal cortex might also be associated with its increasing specialization for visuospatial and attentional functions (Casey et al., 2005).

Visuospatial Processing on Adolescence

In the past several years, there has been an explosive growth in the field of human brain imaging. Much of this work has demonstrated that information processing in the human brain involves the concerted activity of multiple, spatially distributed cortical regions.

Major goals for those engaged in human brain mapping of cognitive function include identifying the regions that participate in various information processing operations, characterizing the functional role played by each region, and modeling the interactions among regions. Vision and visual memory have been especially fruitful domains for studies of distributed neural systems in the human brain (Ungerleider et al. 1998).

The differential visual impairments produced by focal lesions in clinical cases suggest that human visual cortex, like that of the monkey, may be similarly organized into two anatomically distinct and functionally specialized ventral and dorsal processing pathways, occupying about one-half of the total cortex (Desimone and Ungerleider,1989;Felleman and Van Essen, 1991).

These areas are organized into two functionally specialized processing pathways, each having the primary visual cortex (V1) as its source and each being composed of multiple areas beyond V1.

The occipitotemporal pathway, or "ventral stream", is crucial for the identification of objects, whereas the occipitoparietal pathway, or "dorsal stream", is crucial for spatial perception (Ungerleider and Mishkin, 1982; Ungerleider and Haxby, 1994) and for the visual guidance of movements toward objects in space (Goodale and Milner, 1992; Jeannerod and Rossetti, 1993). A simple way to conceptualize the functions of the two processing streams is "what" versus "where".

The ventral stream in the human brain is located almost entirely on the ventral surface of the temporal lobe; in addition, it does not extend as far anteriorly in the human compared with the monkey. The dorsal stream in the human brain has a more superior location in parietal cortex than in the monkey. Such differences suggest that these regions of cortex were displaced in the course of human brain evolution.

The inferior displacement of ventral stream areas and the superior displacement of dorsal stream areas both may be related to the emergence of phylogenetically newer areas specialized for language in the posterior perisylvian cortex. Similarly, the posterior displacement of ventral stream areas away from the temporal pole may be related to the role the latter cortex plays in semantic and lexical knowledge about objects (Damasio et al., 1996; Nobre et al., 1994).

The current version of the visual hierarchy includes 10 levels of cortical processing. Altogether, it contains 14 levels if one includes the retina and lateral geniculate nucleus at the bottom as well as the entorhinal cortex and hippocampus at the top. Within this hierarchy, there are multiple, intertwined processing streams, which, at a low level, are related to the compartimental organization of areas V1 and V2 and, at a high level, are related to the distinction between processing centers in the temporal and parietal lobes (Felleman and Van Essen, 1991).

Sex differences in processing and performance within the visuospatial domain have been of great interest. Generally, tasks involving spatial transformations (i.e., mental rotation) produce the most robust sex-based differences, with males outperforming females; however, differences have also been reported on spatial perception/visualization tasks (Voyer et al., 1995). Most studies have typically addressed the issue of identifying and understanding the neurobiology of sexbased differences using mental rotation tasks as these tend to produce robust results (Butler et al., 2006; Halari et al., 2006; Hugdahl et al., 2006; Jordan et al., 2002; Kucian et al., 2007; Thomsen et al., 2000; Weiss et al., 2003). Additionally, a limited number of studies have utilized more basic perceptual tasks, such as spatial line orientation tasks, to assess whether differences are evident at a more basic level of processing and not only during greater cognitive demands (Clements et al., 2006; Gur et al., 1991, 2000).

A limited number of studies have investigated the neurobiology of sex differences as related to development in visuospatial processing in younger populations. Kucian and colleagues (2007) conducted a study comparing activation associated with a mental rotation task in children and adults. Both groups had activation in right fusiform gyrus and bilateral parietal lobe—superior parietal for the child group (R > L) and inferior parietal for the adult group; adults had additional activation in bilateral extrastriate cortex. The child group showed additional activation when compared to adults, in bilateral posterior cingulate and precuneus while adults had greater activation than children did in left intraparietal sulcus. There were no sex differences in the child group while adult females showed stronger activation in right middle temporal gyrus, right inferior frontal gyrus, and left primary motor. Thus, this study shows that there is a developmental shift in the intraparietal sulcus from stronger right lateralization in younger groups to a bilateral representation in adults.

In spatial line orientation task for children between 7 and 15 years of age, further differences in the development of processing visuospatial information are confirmed through the correlation data. With increasing age, males shift from traditional right hemisphere network (including IPL) to engagement of the left hemisphere; this is consistent with previous literature (Clements et al., 2006; Gur et al., 1991, 2000).These findings suggest that an efficient right hemisphere network is developed at an early age for males and that with increasing age the left hemisphere is recruited and both hemispheres are utilized in a parallel fashion.

Furthermore, these results could indicate that regions within the left hemisphere, for males, are part of an efficient visuospatial and/or visuomotor network. However, a different developmental pattern is seen in females. With increasing age, females shift from the engagement of bilateral medial frontal and posterior regions (e.g., medial frontal gyrus and cingulate) to more traditional areas seen in visuospatial processing (e.g., rIPL, rITG, rDLPFC, bilateral SPL) (Butler et al., 2006; Weiss et al., 2003). Findings suggest that younger females recruit regions typically seen in decision-making and possibly verbal strategy use and as females get older, greater activation is seen in right hemisphere regions. Specifically, activation in females is seen in more traditional areas associated with visuospatial processing, in particular, spatial attention and working memory.

The few studies examining fMRI response during verbal and spatial working memory in children and adolescents suggest that, overall, children and adolescents demonstrate similar frontal and parietal patterns of response as adults (Casey et al., 1995; Thomas et al., 1999), but show greater (Klingsberg et al., 2002; Kwon et al., 2002) and more widespread (Kwon et al., 2002) activation in these regions with increasing age. Schweinsburg et al. (2005) conducted a study investigating the impact of age and sex on spatial working memory. They showed that there was a significant positive correlation with age in a left prefrontal region and bilateral inferior posterior parietal lobe; a significant negative correlation with age was seen in bilateral superior parietal cortex. Moreover, females demonstrated a diminished response in anterior cingulate activation while males demonstrated a greater response in frontopolar cortex compared to females. Schweinsburg et al. concluded that frontal and parietal neural networks that are involved in spatial working memory change during adolescences, influenced by sex, which possibly reflects the development of sex-specific strategies used for task completion.

These neurocognitive studies suggested that brain regions activation during visuospatial processing in adolescence could be correlated with the developmental trajectory of brain neuromaturation processes and also demonstrate the gradual developing of dorsal and ventral visual processing pathways on the brain during adolescence. Also, the sex difference in visuospatial processing could be correlated with the difference in development of brain regions involved in visuospatial abilities under gonadal steroid hormones, testosterone and estrogens, during adolescence.

Conclusion

Studying the human adolescent brain has sparked scientific interest in studying adolescence from a neurobiological perspective, and it has captured the neuroscientist's fascination with the profound neuronal rewiring that takes place during this period of development.

This review highlights the organizational and activational role of steroid hormones, which become elevated during puberty, in the development and remodeling the adolescent brain and cognitive abilities.

The effects of steroid hormones during adolescence occur in both females and males, also across a wide range of species including humans, and influence many behaviors and cognitive abilities such as visuospatial processing.

Examining sex-based differences in visuospatial processing in younger populations could provide further insights into the maturation of important regions involved in visuospatial processing as well as identify potential differences in strategies used, and the development of these strategies under organizational and activational effects of steroid hormones.

One research propose for developmental cognitive neuroscientists could be the neurobiology of cognitive abilities on younger population in visual processing and semantic association between visual information and abstract categories as assessing the neuromaturation of "high level" pathways during adolescence. Direct observations on my young students between 10 and 17 years old, revealed differences between sexes, age and their education. It is possible that gonadal steroid hormones could play an important role in brain regions development involved in visuospatial processing, spatial memory and semantic association during adolescence.

Also, understanding the nature of visuospatial abilities and their relationships to general intelligence or to general-purpose executive functions, should contribute strongly to more fair and comprehensive tests of intelligence (Miyake et al. 2001).

References

Andersen, S. L., Rutstein, M., Benzo, J. M., Hostetter, J. C. & Teicher, M. H. (1997). Sex differences in dopamine receptor overproduction and elimination. *Neuroreport, 8,* 1495-8.

Baddeley, A. (1992). Working memory. *Science*, *255*, 556-559.

Baddeley, A. D. (1986). *Working memory*. Oxford University, Press; Oxford.

Barres, B. A. & Barde, Y. (2000). Neuronal and glial cell biology. *Current Opinion in Biotechnology, 10(5),* 642-648.

Bartzokis, G. (2004). Age-related myelin breakdown: a developmental model of cognitive decline and Alzheimer's disease. *Neurobiology of Aging, 25(1),* 5-18.

Benes, F. M., Turtle, M., Khan, Y. & Farol, P. (1994). Myelination of a key relay zone in the hippocampal formation occurs in the human brain during childhood, adolescence, and adulthood. *Arch Gen Psychiatry, 51,* 477-484.

Blakemore, S. J. & Choudhury, S. (2006). Development of the adolescent brain: implications for executive function and social cognition. *J.Child Psychol Psychiatry, 47,* 296-312.

Bourgeois, J. P., Goldman-Rakic, P. S. & Rakic, P. (1994). Synaptogenesis in the prefrontal cortex of rhesus monkeys. *Cereb Cortex., 4,* 78-96.

Breedlove, S. M. & Hampson, E. (2002). *Sexual differentiation of the brain and behavior.* In: J. B., Becker, A. M., Breedlove, D., Crews, M. M., Carthy, editors. *Behavioral endocrinology*. 2nd edition. Cambridge (MA): Massachusetts Institute of Technology Press, 39-74.

Butler, T., Imperato-McGinley, J., Pan, H., Voyer, D., Cordero, J., Zhu Y. S., Stern, E. & Silbersweig, D. (2006). Sex differences in mental rotation: Top-down versus bottom-up processing. *Neuroimage, 32,* 445-456.

Calvert, G. A. (2001). Cross modal processing in the human brain: insights from functional neuroimaging studies. *Cerebral Cortex, 11(12),* 1110-1123.

Casey, B. J., Cohen, J. D., Jezzard, P., Turner, R., Noll, D. C., Trainor, R. J., Giedd, J., Kaysen, D., Hertz-Pannier, L. & Rapoport, J. L. (1995). Activation of prefrontal cortex in children during a nonspatial working memory task with functional MRI. *Neuroimage, 2,* 221-229.

Casey, B. J., Galvan, A. & Hare, T. A. (2005). Changes in cerebral functional organization during cognitive development. *Curr. Opin. Neurobiol, 15,* 239-244.

Caviness, V. S., Jr, Kennedy, D. N., Richelme, C., Rademacher, J. & Filipek, P. A. (1996). The human brain age 7-11 years: a volumetric analysis based on magnetic resonance images. *Cereb. Cortex. 6*, 726-736.

Chesler, E. J. & Juraska, J. M., (2000). Acute administration of estrogen and progesterone impairs the acquizition of the spatial Morris water maze in ovariectomized rats. *Horm.Behav., 38*, 234-242.

Christiansen, K. & Knussmann, R. (1987). Sex hormone and cognitive functioning in men. *Neuropsychology, 18*, 27-36.

Christiansen, K. (1993). Sex hormone-related variations of cognitive performance in Kung San hunter-gatherers of Namibia. *Neuropsichobiology, 27*, 97-107.

Chung, W. C., De Vries, G. J. & Swaab, D. F. (2002).Sexual differentiation of the bed nucleus of the stria terminalis in humans may extend into adulthood. *Journal of Neuroscience, 22*, 1027-33.

Clements, A. M., Rimrodt, S. L., Abel, J. R., Blankner, J. G., Mostofsky, S. H., Pekar, J. J., Denckla, M. B. & Cutting, L. E. (2006). Sex differences in cerebral laterality of language and visuospatial processing. *Brain Lang., 98*, 150-158.

Cooke, B., Hegstrom, C. D., Villeneuve, L. S. & Breedlove, S. M. (1998). Sexual differentiation of the vertebrate brain: principles and mechanisms. *Front Neuroendocrinol, 19*, 323-362.

Damasio, H., Grabowski, T. J., Tranel, D., Hichwa, R. D. & Damasio, A. R. (1996) A neural basis for lexical retrieval. *Nature, 380*, 499-505.

Daniel, J. M., Fader, A. J., Spencer, A. & Dohanich, G. P. (1997). Estrogen enhances performance of female rats during acquisition of a radial arm maze. *Horm.Behav., 32*, 217-225.

Davis, E. C., Shryne, J. E. & Gorski, R. A. (1996). Structural sexual dimorphisms in the anteroventral periventricular nucleus of the rat hypothalamus are sensitive to gonadal steroids perinatally, but develop peripubertally. *Neuroendocrinology, 63*, 142-8.

De Bellis, M. D., Keshavan, M. S., Beers, S. R., Hall, J., Frustaci, K., Masalehdan, A., Noll, J. & Boring, A. M. (2001). Sex differences in brain maturation during childhood and adolescence. *Cereb Cortex., 11*, 552-557.

Dekaban, A. S. (1978). Changes in brain weights during the span of human life: relation of brain weights to body heights and body weights. *Annals of Neurology, 4*, 345-56.

Desimone, R. & Ungerleider, L. G. (1989). in *Handbook of Neuropsychology*, eds. Boller, F. and Grafman, J.(Elsevier, Amsterdam), Vol. *2*, 267-300.

Du, Y. & Dreyfus, C. F. (2002). Oligodendrocytes as providers of growth factors. *Journal of Neuroscience Research, 68(6)*, 647-654.

Eckenhoff, M. F. & Rakic, P. (1988). Nature and fate of proliferative cells in the hippocampal dentate gyrus during the life span of the rhesus monkey. *Journal of Neuroscience, 8*, 2729-47.

Errico, A. L., Parsons, O. A., Kling, O. R. & King, A. C. (1992). Investigation of the role of sex hormones on alchoolics visuospatial deficits. *Neuropsychologia, 30*, 417-426.

Felleman, D. J. & Van Essen, D. C. (1991). Distributed hierarchical processing in the primate cerebral cortex Cereb. Cortex., *1*, 1-47.

Fields, R. D. & Stevens-Graham, B. (2002). New insights into neuron-glia communication. *Science, 298(5593)*, 556-562.

Fields, R. D., Eshete, F., Dudek, S., Ozsarac, N. & Stevens, B. (2001). Regulation of gene expression by action potentials, dependence on complexity in cellular information processing. *Novartis Foundation Symposium, 239*, 160-172 (discussion 172-6, 234-40).

Fuster, J. M. (1989). *The prefrontal cortex*. New York: Raven.

Galea, L. A., Spritzer, M. D., Barker, J. M. & Pawluski, J. L. (2006). Gonadal hormone modulation of hippocampal neurogenesis in the adult. *Hippocampus, 16*, 225-232.

Galvan, A., Hare, T., Voss, H., Glover, G. & Casey, B. J. (2007). Risk-taking in the adolescent brain: who is at risk? *Dev. Sci., 10*, F8-F14.

Gathercole, S. E. (1999). Cognitive approaches to the development of short-term memory. *Trends in Cognitive Sciences, 3,* 410-419.

Gathercole, S. E., Pickering, S. J., Ambridge, B. & Wearing, H. (2004).The structure of working memory from 4 to 15 years of age. *Developmental Psychology, 40,* 177-190.

Giedd, J. N., Blumenthal, J., Jeffries, N. O., Castellanos, F. X., Liu, H., Zijdenbos, A., Paus, T., Evans, A. C. & Rapoport , J. L. (1999). Brain development during childhood and adolescence: a longitudinal MRI study. *Nat Neurosci. 2,* 861-863.

Giedd, J. N., Blumenthal, J., Jeffries, N. O., Castellanos, F. X., Liu, H., Zijdenbos, A., Paus, T., Evans, A. C. & Rapoport, J. L. (1999b). Brain development during childhood and adolescence: a longitudinal MRI study. *Nature Neuroscience, 2(10)*, 861-863.

Giedd, J. N., Castellanos, F. X., Rajapakse, J. C., Vaituzis, A. C. & Rapoport, J. L. (1997). Sexual dimorphism of the developing human brain. *Prog Neuropsychopharmacol. Biol.Psychiatry., 21*, 1185-1201.

Giedd, J. N., Clasen, L. S., Lenroot, R., Greenstein, D., Wallace, G. L., Ordaz, S., Molloy, E. A., Blumenthal, J. D., Tossell, J. W., Stayer, C., et al. (2006). Pubertyrelated influences on brain development. *Mol Cell Endocrinol. 25,* 154-162.

Gilmore, J. H., Lin, W., Prastawa, M. W., Looney, C. B., Vetsa, Y. S., Knickmeyer, R. C., Evans, D. D., Smith, J. K., Hamer, R. M., Lieberman, J. A., et al. (2007). Regional gray matter growth, sexual dimorphism, and cerebral asymmetry in the neonatal brain. *J Neurosci., 27*, 1255-1260.

Gogtay, N., Giedd, J. N., Lusk, L., Hayashi, K. M., Greenstein, D., Vaituzis, A. C., Nugent, T. F., 3rd, Herman, D. H., Clasen, L. S., Rapoport, J. L., et al. (2004). Dynamic mapping of human cortical development during childhood through early adulthood. *Proc Natl Acad Sci USA., 101*, 8174-8179.

Goldstein, J. M., Kennedy, D. N. & Caviness, V. S. Jr. (1999). Images in neuroscience. Brain development 11, sexual dimorphism. *Am J Psychiatry. 156*, 352.

Goodale, M. A. & Milner, A. D. (1992). Separate visual pathways for perception and action. *Trends Neurosci. 15*, 20-25.

Gur, R. C., Alsop, D., Glahn, D., Petty, R., Swanson, C. L., Maldjian, J. A., Turetsky, B. I., Detre, J. A., Gee, J. & Gur, R. E. (2000). An fMRI study of sex differences in regional activation to a verbal and a spatial task. *Brain Lang., 74*, 157-170.

Gur, R. C., Raglan, J. D., Resnick, S. M., Skolnick, B. E., Jaggi, J., Muenz, L. & Gur, R. E. (1991). Lateralized increases in cerebral blood flow during performance of verbal and spatial tasks: relationship with performance level. *Brain Cogn., 24,* 244-258.

Gur, R. C., Turetsky, B. I., Matsui, M., Yan, M., Bilker, W. & Hughett, P., Gur, R. E. (1999). Sex differences in brain gray and white matter in healthy young adults: correlations with cognitive performance. *J Neurosci. 19*, 4065-4072.

Gurzu, C., Artenie, V., Hritcu, L. & Ciobica, A. (2008). Prenatal testosterone improves the spatial learning and memory by protein synthesis in different lobes of the brain in the male and female rat. *Cent. Eur. J. Biol., 3(1)*, 39-47.

Halari, R., Sharma, T., Hines, M., Andrew, C., Simmons, A. & Kumari, V. (2006). Comparable fMRI activity with differential behavioural performance in mental rotation and overt verbal fluency tasks in healthy men and women. *Exp. Brain Res., 169*, 1-14.

He, J. & Crews, F. T. (2007). Neurogenesis decreases during brain maturation from adolescence to adulthood. *Pharmacology, Biochemistry & Behavior, 86*, 327-33.

Hugdahl, K., Thomsen, T. & Ersland, L. (2006). Sex differences in visuo-spatial processing: an fMRI study of mental rotation. *Neuropsychologia, 44*, 1575-1583.

Huttenlocher, P. R. & Dabholkar, A. S. (1997). Regional differences in synaptogenesis in human cerebral cortex. *Journal of Comparative Neurology, 387*, 167-78.

Huttenlocher, P. R. & de Courten, C. (1987). The development of synapses in striate cortex of man. *Hum Neurobiol., 6*, 1-9.

Janovsky, J. S., Oviatt, S. K. & Orwoll, E. S. (1994). Testosterone influences spatial cognition in older men. *Behav.Neurosci,108*, 325-332.

Jeannerod, M. & Rossetti, Y. (1993). Visuomotor coordination as a dissociable visual function: experimental and clinical evidence. *Bailliere Clin. Neurol., 2*, 439-460.

Jernigan, T. L., Trauner, D. A., Hesselink, J. R. & Tallal, P. A. (1991).Maturation of human cerebrum observed in vivo during adolescence.*Brain,114,* 2037-2049.

Jordan, K., Wüstenberg, T., Heinze, H. J., Peters, M. & Jäncke, L. (2002).Women and men exhibit different cortical activation patterns during mental rotation tasks. *Neuropsychologia, 40*, 2397-2408.

Kinney, H. C., Karthigasan, J., Borenshteyn, N. I., Flax, J. D. & Kirschner, D. A. (1994). Myelination in the developing human brain: biochemical correlates. *Neurochemical Research, 19(8)*, 983-996.

Klingberg, T., Forssberg, H. & Westerberg, H. (2002). Increased brain activity in frontal and parietal cortex underlies the development of visuospatial working memory capacity during childhood. *Journal of Cognitive Neuroscience., 14,* 1-10.

Kucian, K., von Aster, M., Loenneker, T., Dietrich, T., Mast, F. W. & Martin, E. (2007). Brain activation during mental rotation in school children and adults. *J. Neural Transm., 114*, 675-686.

Kwon, H., Reiss, A. L. & Menon, V. (2002). Neural basis of protracted developmental changes in visuo-spatial working memory. *Proceedings of the National Academy of Sciences, U.S.A., 99,* 13336-13341.

Lange, N., Giedd, J. N., Castellanos, F. X., Vaituzis, A. C. & Rapoport, J. L. (1997). Variability of human brain structure size, ages 4-20 years. *Psychiatry Res., 74*, 1-12.

Lenroot, R. K. & Giedd, J. N. (2006). Brain development in children and adolescents: insights from anatomical magnetic resonance imaging. *Neuroscience & Biobehavioral Review, 30*, 718-29.

Lenroot, R. K., Gogtay, N., Greenstein, D. K., Wells, E. M., Wallace, G. L., Clasen, L. S., Blumenthal, J. D., Lerch, J., Zijdenbos, A. P., Evans, A. C., et al. (2007). Sexual dimorphism of brain development trajectories during childhood and adolescence. *Neuroimage, 36*, 1065-1073.

Leranth, C., Petnehazy, O. & MacLusky, N. J. (2003). Gonadal hormones affect soine synaptic density in the CA1 hippocampal subfield of male rats. *J.Neurosci, 23(5)*, 1588-1592.

Leranth, C., Prange-Kiel, J., Frick, K. M. & Horvarth, T. L. (2004). Low CA1 spine synapse density is furtther reduced by castration in male non-human primates. *Cereb-Cortex., 14(5)*, 503-510.

Luna, B., Garver, K. E., Urban, T. A., Lazar, N. A. & Sweeney, J. A. (2004). Maturation of cognitive processes from late childhood to adulthood. *Child Development, 75,* 1357-1372.

MacLusky, N. J., Hajszan, T., Prange-Kiel, J. & Leranth, C. (2006). Androgen modulation of hippocampal synaptic plasticity. *Neuroscience, 138,* 957-965.

Markham, J. A., Morris, J. R. & Juraska, J. M. (2007). Neuron number decreases in the rat ventral, but not dorsal, medial prefrontal cortex between adolescence and adulthood. *Neuroscience, 144*, 961-968.

Martin, A. & Chao, L. L. (2001). Semantic memory and the brain: structure and processes. *Current Opinion in Neurobiology, 11(2)*, 194-201.*Med. Sci.Monit., 6*, 1186-1196.

Mesulam, M. M. (1998). From sensation to cognition. *Brain 121 (Part 6)*, 1013-1052.

Miller, E. K. & Cohen, J. D. (2001).An integrative theory of prefrontal function. *A. Rev. Neurosci., 24*, 167-202.

Miyake, A., Friedman, N. P., Rettinger, D. A., Shah, P. & Hegarty, M. (2001). How Are Visuospatial Working Memory, Executive Functioning, and Spatial Abilities Related? A Latent-Variable Analysis. *Journal of Experimental Psychology, General, 130(4)*, 621-540.

Naghdi, N., Nafisy, N. & Majlessi, N. (2001). The effects of intrahippocampal testosterone and flutamide on spatial localization in the Morris water maze. *Brain Res., 897*, 44-51.

Nass, R. & Baker, S. (1991). Androgen effects on cognition: Congenital adrenal hyperplasia. *Psychoneuroendocrinology, 16,* 189-201.

Nieder, A., Freedman, D. J. & Miller, E. K. (2002). Representation of the quantity of visual items in the primate prefrontal cortex. *Science, 297*, 1708-1711.

Nobre, A. C., Allison, T. & McCarthy, G. (1994). Word recognition in the human inferior temporal lobe. *Nature, 372*, 260-263.

Nunez, J. L., Lauschke, D. M. & Juraska, J. M. (2001).Cell death in the development of the posterior cortex in male and female rats. *Journal of Comparative Neurology, 436*, 32-41.

Nunez, J. L., Sodhi, J. & Juraska, J. M. (2002). Ovarian hormones after postnatal day 20 reduce neuron number in the rat primary visual cortex. *Journal of Neurobiology, 52*, 312-21.

Paus, T., Collins, D. L., Evans, A. C., Leonard, G., Pike, B. & Zijdenbos, A. (2001). Maturation of white matter in the human brain: a review of magnetic resonance studies. *Brain Research Bulletin, 54*, 255-66.

Phillips, S. M. & Sherwin, B. B. (1992a). Effects of estrogen on memory function in surgically menopausal women. *Psychoneuroendocrinology, 17*, 485-495.

Pinos, H., Collado, P., Rodriguez-Zafra, M., Rodriguez, C., Segovia, S. & Guillamon, A. (2001). The development of sex differences in the locus coeruleus of the rat. *Brain Research Bulletin, 56*, 73-8.

Postma, A., Meyer, G., Tuiten, A., van Honk, J., Kessels, R. P. & Thijssen, J. (2000). Effects of testosterone administration on selective aspects of object-location memory in healthy young women. *Psychoneuroendocrinology, 25*, 562-575.

Purves, D., Brannon, E., Cabeza, R., Huettel, A. S., LaBar, K., Platt, M. & Woldorff, M. G. (2008). *Principles of Cognitive Neuroscience*, Sinauer Associates Inc.Sunderland, Massachusetts, USA.

Rankin, S. L., Partlow, G. D., McCurdy, R. D., Giles, E. D. & Fisher K. R. (2003). Postnatal neurogenesis in the vasopressin and oxytocin-containing nucleus of the pig hypothalamus. *Brain Research, 971*, 189-96.

Romeo, R. D., Richardson, H. N. & Sisk, C. L. (2002). Puberty and the maturation of the male brain and sexual behavior: recasting a behavioral potential. *Neurosci Biobehav Rev., 26*, 381-391.

Schmithorst, V. J., Holland, S. K. & Dardzinski, B. J. (2007). Developmental differences in white matter architecture between boys and girls. *Human Brain Mapp., 29*, 696-710.

Schweinsburg, A. D., Nagel, B. J. & Tapert, S. F. (2005). fMRI reveals alteration of spatial working memory networks across adolescence. *J. Int. Neuropsychol. Soc., 11*, 631-644.

Sherwin, B. B. (1994). Estrogenic effects on memory in women. *Ann.N.Y.Acad.Sci., 743*, 213-232.

Simerly, R. B., Chang, C. & Muramatsu, M. & Swanson, L. W. (1990). Distribution of androgen and estrogen receptor mRNA-containing cells in the rat brain: an in situ hybridization study. J Comp Neurol., *294(1),* 76-95.

Sisk, C. L. & Foster, D. L. (2004). The neural basis of puberty and adolescence. *Nat Neurosci., 7,* 1040-1047.

Sowell, E. R., Delis, D., Stiles, J. & Jernigan, T. L. (2001). Improved memory functioning and frontal lobe maturation between childhood and adolescence: a structural MRI study. *J Int Neuropsychol Soc, 7*, 312-322.

Sowell, E. R., Peterson, B. S., Kan, E., Woods, R. P., Yoshii, J., Bansal, R., Xu, D., Zhu, H., Thompson, P. M. & Toga, A. W. (2007). Sex differences in cortical thickness mapped in 176 healthy individuals between 7 and 87 years of age. *Cereb Cortex., 17(7)*, 1550-1560.

Sowell, E. R., Thompson, P. M., Leonard, C. M., Welcome, S. E., Kan, E. & Toga, A. W. (2004). Longitudinal mapping of cortical thickness and brain growth in normal children. *Journal of Neuroscience, 24*, 8223-31.

Sowell, E. R., Thompson, P. M., Tessner, K. D. & Toga, A. W. (2001). Mapping continued brain growth and gray matter density reduction in dorsal frontal cortex: Inverse relationships during postadolescent brain maturation. *Journal of Neuroscience, 21*, 8819-29.

Sowell, E. R., Trauner, D. A., Gamst, A. & Jernigan, T. L. (2002). Development of cortical and subcortical brain structures in childhood and adolescence: a structural MRI study. *Dev Med Child Neurol., 44*, 4-16.

Tanner, J. M. & Davies, P. S. (1985). Clinical longitudinal standards for height and height velocity for North American children. *J Pediatr. 107,* 317-329.

Thomas, K. M., King, S. W., Franzen, P. L., Welsh, T. F., Berkowitz, A. L., Noll, D. C., Birmaher, V. & Casey, B. J. (1999). A developmental functional MRI study of spatial working memory. *Neuroimage, 10*, 327-338.

Thomsen, T., Hugdahl, K., Ersland, L., Barndon, R., Lundervold, A., Smievoll, A. I., Roscher, B. E. & Sundberg, H. (2000). Functional magnetic resonance imaging (fMRI) study of sex differences in a mental rotation task.

Ungerleider, L. G. & Haxby, J. V. (1994). "What" and "where" in the human brain. *Current Opinion in Neurobiology, 4*, 157-165.

Ungerleider, L. G. & Mishkin, M. (1982). in *Analysis of Visual Behavior*, eds. D. J., Ingle, M. A. Goodale, & R. J. Mansfield, W.MIT Press, Cambridge, MA, 549-586.

Ungerleider, L. G., Courtney, S. M. & Haxby, J. V. (1998). A neural system for human visual working memory, *Proc. Natl. Acad. Sci. USA, 95,* 883-890.

van Goozen, S. H., Cohen-Kettenis, P. T., Gooren, L. J., Frijda, N. H. & van den Poll, N. E. (1995). Gender differences in behavior:Activating effects of cross-sex hormones. *Psychoneuroendocrinology, 20*, 343-363.

Volpe, J. J. (2000). Overview: normal and abnormal human brain development. Mental Retardation and Developmental Disabilities *Research Reviews, 6(1)*, 1-5.

Voyer, D., Voyer, S. & Bryden, M. P. (1995). Magnitude of sex differences in spatial abilities: A meta-analysis and consideration of critical variables. *Psychol.Bull., 117*, 250-270.

Vuontela, V., Steenari, M. R., Carlson, S., Koivisto, J., Fjallberg, M. & Aronen, E. T. (2003). Audiospatial and visuospatial working memory in 6−13 year old school children. *Learning and Memory, 10,* 74-81.

Weiss, E., Siedentopf, C. M., Hofer, A., Deisenhammer, E. A., Hoptman, M. J., Kremser, C., Goalszewski, S., Felber, S., Fleischhacker, W. W. & Delazer, M. (2003). Sex differences in brain activation pattern during a visuospatial cognitive task: A functional magnetic resonance imaging study in healthy volunteers. *Neurosci. Lett., 344*, 169-172.

Wilke, M., Krageloh-Mann, I., Holland, S. K. (2007). Global and local development of gray and white matter volume in normal children and adolescents. *Exp Brain Res., 178*, 296-307.

Wooley, C. S. & McEwen, B. (1994). Estradiol regulates hippocampal dendritic spine density via an N-methyl-D-aspartate receptor-dependent mechanism. *J.Neurosci, 14*, 7680-7687.

Yakovlev, P. I. & Lecours, A. (1967). The myelogenetic cycles of regional maturation of the brain. In: Minkovski, A. (Ed.), *Regional development of the brain in early life*. Blackwell, Oxford 3-65.

Zald, D. H. & Iacono, W. G. (1998). The development of spatial working memory abilities. *Developmental Neuropsychology, 14,* 563-578.

In: Spatial Memory ISBN: 978-1-61668-139-5
Editor: Jacob B. Thomas, pp. 123-144

Chapter 5

FAMILIARITY AND SPATIAL COGNITIVE STYLE: HOW IMPORTANT ARE THEY FOR SPATIAL REPRESENTATION?

Raffaella Nori[1*] and Laura Piccardi[2,3]
[1]Dipartimento di Psicologia, Università degli Studi di Bologna
[2]Dipartimento di Scienze della Salute, Università degli Studi di L'Aquila
[3]Centro Ricerche di Neuropsicologia, I.R.C.C.S., Fondazione Santa Lucia, Roma

ABSTRACT

How we acquire and represent spatial information is one of the most important unsolved issues in spatial cognition. Siegel and White (1975) affirmed that different forms of environmental knowledge are acquired and represented depending on the type of information selected: *landmark*, characterized by environmental patterns that are perceptually salient or important for the person; *route*, based on the pathes generally used to connect landmarks; and *survey*, an overall configuration of the environment, similar to a map. According to these authors, anyone can reach survey representation with extensive experience of the environment. Recently, Pazzaglia and co-workers (2000) demonstrated

* Corresponding author: Department of Psychology, University of Bologna, V.le Berti Pichat, 5, 40127 Bologna, Italy, Phone number: +39 051 2091885m Fax: +39 051 243086, E-mail: raffaella.nori@unibo.it

that these three types of representations correspond to three different spatial cognitive styles. Conversely, according to Montello (1998) a pure landmark or route representation does not exist; indeed, during their first exposure to the environment, people acquire and represent an overall survey configuration of it.

Our aim was to determine whether environmental familiarity and/or spatial cognitive style predict the way we acquire and represent spatial information.

Forty participants who had different degrees of familiarity with the Italian city of Bologna took part in the experiment. Familiarity with Bologna was evaluated using a questionnaire. This city was selected because it has a small and well-defined centre that can be easily explored on foot. Participants were further subdivided by spatial cognitive style to assess its weight in environmental representation. They performed six spatial tasks concerning Bologna that measured different spatial abilities, based on Siegel and White's frameworks (1975).

We found that neither familiarity with the environment nor spatial cognitive style predict the correct solution of landmark tasks, whereas both familiarity with the environment and spatial cognitive style predict the correct solution of route and survey tasks.. Thus, we can affirm that both familiarity with the environment and spatial cognitive style are important for acquiring and representing spatial information, but their involvement depends on task demands. To our knowledge, this is the first study in which knowledge of a real town has been measured. This ecological setting allowed us to propose a new model to explain individual differences in moving successfully through the environment.

INTRODUCTION

Few everyday life problems are more fundamental than getting from here to there. Humans and other animals move about their environments in order to get to places with food or shelter and other resources; they also have to avoid threats and dangers such as predator, assaults and other risk factors. Although navigation is a very important activity in daily life, it is still not clear how we acquire and mentally represent the environment. In an influential paper, Siegel and White (1975) argued that the first acquirement and representation of the environment is *landmark*, then *route* and finally *survey*. Specifically, "*landmark* representation" is characterized by environmental patterns that are perceptually salient or important for people, such as their home. This kind of spatial representation does not provide any spatial information. The prominent role of landmarks in spatial representation seems to require a special kind of figurative memory.

"*Route* representation" is based on landmarks and on the routes generally used to connect those landmarks. While landmark representation is considered visual, route representation is predominantly sensorimotor. In fact, a person memorizes routes or paths like a mental list of distances and directions that must be followed in precise sequences of motor actions. This representation is organized on the basis of body references, that is, an egocentric frame of references or coordinates. It is characterized by sequential rather than spatial aspects. Each element of the sequence is useful because it follows one element and leads to another one.

Finally, "*survey* representation" is an overall configuration of the environment. It implies the encoding of directions and distances between places regardless of the person's position. It is based on allocentric frames of references or coordinates (e.g., cardinal points) and characterized by a high level of plasticity.

This model is hierarchical, because landmark representation is characterized only by the properties of landmark representation; route representation is characterized by the properties of both landmark and route representation; and survey representation is characterized by the proprieties of landmark, route, and survey representation. According to Siegel and White (1975), anyone can reach survey representation through extensive experience in the environment.

Recently, Pazzaglia, Cornoldi, and De Beni (2000) and Pazzaglia and De Beni (2001) demonstrated that these three types of mental representation correspond to three different cognitive styles. In Pazzaglia and co-workers (2000), regardless of their ability to navigate a number of participants reported that they preferred landmark representation as their usual approach for moving in the environment (Denis, Pazzaglia, Cornoldi, & Bertolo, 1999). Instead, people who prefer to use patterns that are perceptually salient or important for them (landmarks) and egocentric coordinates are considered route style users. Finally, people who use both patterns that are perceptually salient for them and egocentric and allocentric coordinates are considered survey style users. Unlike Siegel and White's model, this theoretical approach (Pazzaglia et al., 2000; Pazzaglia & De Beni, 2001) hypothesizes that a person can stop at a specific spatial representation without ever reaching the others. Although both models are hierarchical, the first one predicts that anyone can reach survey representation through extensive experience of the environment and the second one that a person might stop at a particular spatial representation which corresponds to his preferred spatial cognitive style (Siegel & White, 1975; Pazzaglia et al., 2000).

According to Montello (1998), one particularly unconvincing aspect of Siegel & White's model is that landmark representation is a necessary prerequisite for route representation, which, in turn, is a necessary prerequisite for survey representation. Indeed, several studies have shown that following minimal navigation through a new environment individuals are able to perform tasks that require survey knowledge, that is, tasks such as taking shortcuts and estimating distances or giving directions (Klatzky, Loomis, Golledge, Cicinelli, Doherty, & Pellegrino, 1990; Landau, Spelke, & Gleitman, 1984; Loomis, Klatzky, Golledge, Cicinelli, Pellegrino, & Fray, 1993).

In response to these issues, Montello (1998) proposed an alternative framework to explain how people acquire mental representations of space.

He suggested five major tenets:

1. There is no stage at which pure landmark or route knowledge exists, but metric configurational knowledge begins to accrue on first exposure to a novel place;
2. With additional locomotor and perceptual experience of the environment, familiarity with the place increases, resulting in more extensive, complete, and accurate knowledge of the place;
3. A qualitative shift in environmental knowledge results from the integration of knowledge about separately learned places. That is, spatial knowledge about an environment experienced during on separate occasions is stored together in memory. This integration transforms spatial knowledge into more complex, hierarchically organized knowledge;
4. People with equal levels of exposure to an environment differ in the extent and accuracy of their spatial knowledge;
5. Finally, linguistic systems of storing and communicating spatial knowledge provide for the existence of pure topological knowledge, or at least non metric knowledge. However, non metric knowledge exists in addition to metric spatial knowledge but not as a precursor or intrinsic part of it.

In his model, Montello also posited that individual differences, especially in the ability to integrate separately learned places, were so large that they deserved special note. Otherwise, they could confound general statements about the acquisition of spatial mental representation (Montello, 1998). For instance, there could be many individual differences related to metric information (i.e., distance estimation), how it is acquired, how precise it is and whether integration spatial information occurs. In particular, gender differences seem to account partially for the contrasting results reported on these topics (Miller & Santoni, 1986; Ward, Newcombe, & Overton, 1986).

Only few studies have experimentally examined changes in the accuracy of people's spatial knowledge over time (Gärling, Böök, Lindberg, & Nilsson, 1981; Herman, Blomquist, & Klein, 1987; Ishikawa & Montello, 2006; Iachini, Ruotolo, & Ruggiero, 2009). In the studies by Gärling and co-workers (1981) and Herman and co-workers (1987) participants travelled a route. Once they had learned it, they were asked to estimate distance and direction. Participants repeated this set of route travel and estimation tasks several times. Adult participants acquired landmark representation immediately and they improved their direction and distance estimates in the early sessions. The authors never observed perfect accuracy in the participants' performances. These results support Montello's framework.

In an extensive longitudinal study, Ishikawa and Montello (2006) examined 24 college students individually as they drove along two routes in a previously unfamiliar neighbourhood over 10 weekly sessions. Their results contrasted with the predictions of Montello's framework and Siegel and White's model. In fact, although some participants' knowledge improved fairly continuously over the sessions, as hypothesised by Siegel and White's model, most participants either manifested accurate metric knowledge from the first exposure or never manifested accurate metric knowledge, in agreement with Montello's framework. Therefore, they interpreted these results on the basis of large individual differences (Ishikawa & Montello, 2006).

In a recent paper, Iachini and co-workers (2009) reported a study on how familiarity and gender influence the frame of reference used to represent a regular real-world environment, in this case an administrative district of Naples, Italy. Familiar and unfamiliar participants with the experimental environment were asked to learn the location of three triads of buildings by walking guided by examiner along a pathway. Participants included individuals who were either familiar or unfamiliar with the location of the three triads of buildings. Unfamiliar participants had to learn the location of the triads by walking along a pathway guided by the examiner. Then, they had to judge whether the previously learned triad correctly represented the relative positions among the buildings. They found that the accuracy of the unfamiliar participants decreased as much as angular size deviated from 0°. By contrast, the familiar participants were more accurate at 0°-90°-180°. The authors suggest that when the environment is familiar people use allocentric frames of reference, as suggested by Siegel and White's model. Regarding the effect of gender, they observed that males were more accurate and faster than females and that this difference was particularly evident in participants who were unfamiliar with the environment (Iachini et al., 2009).

Except for the few studies discussed above, the acquisition of spatial representation has never been investigated in an ecological way. In fact, learning the environment has always been experimentally guided and, to our knowledge, no studies have explored its natural acquisition in daily life navigation of the person's environment. For this reason, our aim was to determine whether environmental familiarity, as suggested by Siegel and White's model, spatial cognitive style, as suggested by Pazzaglia and co-workers' model, and/or individual differences (i.e., sex), as hypothesised by Montello's model, predict the way we acquire and represent spatial information in a real town.

METHOD

Participants

Forty college students from the Department of Psychology at the University of Bologna who had different degrees of familiarity with the city took part in the experiment: 28 females and 12 males, whose ages ranged from 19 to 28 years (M = 23.97, SD = 1.84). All participants were right-handed according to self-report (Salmaso & Longoni, 1985). Familiarity with Bologna was evaluated using a questionnaire that considered months of stay in Bologna, ranging from 1 to 348 months (M = 140.88, SD = 142.26). All participants gave informed written consent before participating in the experiment.

Experimental Setting and Materials

We investigated spatial knowledge by using a real town, Bologna, selected for its characteristics. In fact, this town has a small, well-defined centre that can be easily explored on foot. Our participants adopted "free exploration" to move around the town. Even if the participants were familiar with the different buildings, monuments and routes in Bologna, we did not ask them to pay particular attention to them or to their location, because their displacement behaviour was not experimentally goal-directed. They learned the environment by means of daily life exploration without any experimental condition restraints.

Preliminary Phase. To be sure that the participants knew the town by navigating through it, they had to complete a questionnaire about Familiarity and Spatial Cognitive Style, which contained questions about how many months they had been living in Bologna and their knowledge of the different parts of the city which we will consider in the experimental section. After analyzing their self-reports, we eliminated two participants because they only moved through a specific and restricted area of the city (i.e., the railway station) and only knew the other parts of Bologna from studying a map and not actually moving through it. Moreover, to evaluate individual spatial cognitive style we modified some questions deriving from Pazzaglia and co-workers (2000) and included them in our questionnaire. In order to label individual spatial cognitive style as landmark, route or survey style, we adopted the method suggested by Nori and Giusberti (2003).

To be labelled as "landmark", participants' score on the questionnaire had to reflect an 80% preference for the landmark cognitive style. Moreover, their preferences on the route and survey questions had to be less than 50%.

Furthermore, in order for participants to be labelled as "route" their score on the questionnaire had to reflect an 80% preference for landmark and route cognitive style and their preferences on the survey questions had to be 50% or less.

Finally, participants, who demonstrated at least an 80% preference for landmark, route and survey questions were labelled as survey style.

Each participant was clearly classifiable in one of the three cognitive styles; no participant obtained the same score in two or more cognitive styles.

Subsequently, participants were submitted to two spatial tasks to evaluate whether they had normal spatial ability. Specifically, to test participants' basic spatial ability, that is, to recognize landmarks and right and left discrimination, we respectively used the *Photo Task* (Nori & Giusberti, 2003, 2006) and the *Money Road Map Test* (Money, Alexander, & Walker, 1965). In the *Photo Task*, participants were asked to choose the photo (target) they had previously studied for 3 seconds, out of four subsequently introduced (i.e., one target, three fillers). The fillers were similar buildings characterized by the presence or absence of specific objects such as a flowerpot, a car or a road sign. Each participant completed seven trials in this task. All responded correctly on at least 6 items, which is within the norm for this task (Nori & Giusberti, 2006). The Money Road Map Test (Money et al., 1965) consists of an A4-sized paper map on which a pathway is traced. Participants are required to imagine they are moving along the pathway and to indicate verbally whether they should make a left or a right turn at each change of direction. Unlike the original version, we allowed the participants to turn the map or to make head and body movements to give the correct answer in order to avoid mental rotation, which is a complex spatial ability (e.g., Nori & Giusberti, 2006; Pazzaglia & De Beni, 2006). This kind of exploration provides a within-map viewpoint, which is an egocentric reference system. Each participant gave at least 23 out of 30 correct responses, which are within in the normal range (Vingerhoets, Lannoo, & Bauwens, 1996).

Experimental phase. Subsequently, in the experimental phase, the participants had to solve 6 spatial tasks that measured different spatial abilities, based on Siegel and White's framework (1975).

Landmark Tasks

1. *Landmark Production Task* (e.g., Giusberti, Nori, & Mercuri, 2009). Participants were simply asked to list 10 landmarks of Bologna. Response time was recorded.

2. *Landmark Recognition Task.* Participants were asked to recognize 10 photos of Bologna's landmarks out of 20 photos (i.e., 10 targets and 10 fillers, that is, landmarks of other Italian cities with architectural features similar to those of Bologna; number of correct responses from 0 to 10) (see figure 1). The order of targets and fillers was randomised. Latency of response was recorded from the time the photo was presented until the participant answered.

A. B.

Figure 1. Examples of two landmarks (A. target and B. filler) used in the Landmark Recognition Task.

These tasks were used to evaluate landmark mental representation ability, because in order to solve them correctly participants had to represent perceptually salient patterns mentally without referring to any kind of spatial information about them.

Route Tasks

1. *Path Task* (based on Nori & Giusberti, 2006). On an A4 sheet of paper, participants viewed two pictures (starting position and goal) of Bologna depicting two well-known landmarks. Then, they were asked to imagine going from the starting position to the goal by walking along on a specific path indicated by the experimenter. Three or four landmarks the person would meet along the path were printed on the same sheet of paper. The participants' task was to arrange the landmarks in the right order according to the previous marked path (Figure 2).
   Participants had to solve 10 trials (number of correct responses from 0 to 10). Latency of response was recorded on each trial from the time when participants saw the landmarks until they responded;

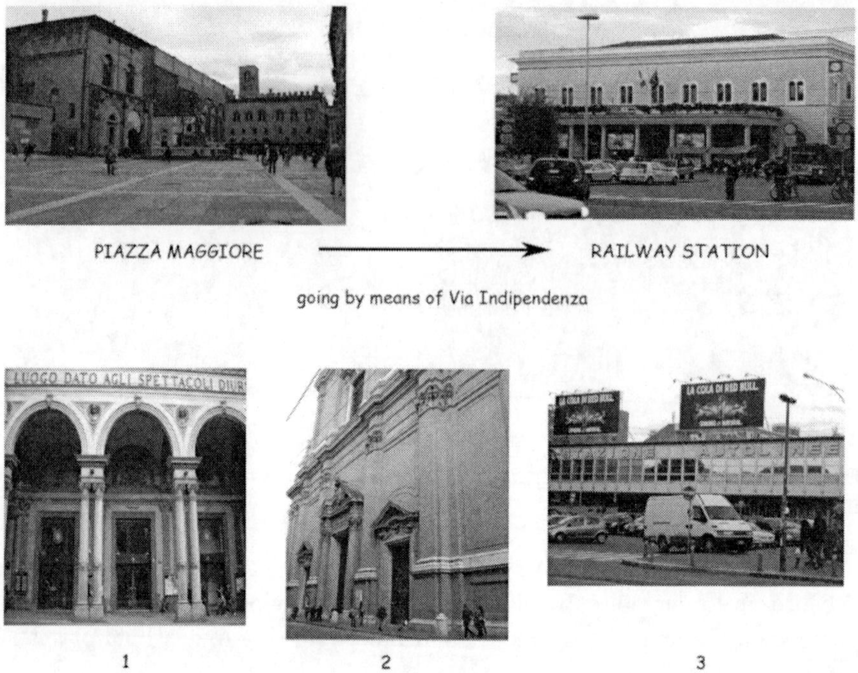

Figure 2. Example of a trial in the Path Task.

2. *Path Description* (based on Denis, et al., 1999). Participants were given instructions similar to those of Denis and co-workers (1999): "Image that a person who is in Bologna for the first time asks you how to go from a particular place to another one on foot, and that you must tell him/her how to do this. You have to describe the path from the starting point to the ending point as accurately as possible, indicating the landmarks along the path and the turns to the left or right. You must do this for 5 routes, giving information you consider to be important for the person to reach the desired goal." All five routes were selected as variants of typical itineraries in the city of Bologna and all connected the starting point and the ending point fairly directly; however, they all deviated from a simple straight line. Route 1 connected Piazza del Nettuno to the Railway Station (1.5 km), Route 2 connected Piazza Verdi to Due Torri (380 m), Route 3 connected San Vitale to Santa Lucia (900 m), Route 4 connected Piazza Maggiore to Piazza Santo Stefano (540 m), and finally Route 5

connected Piazza Scaravilli to Parco della Montagnola (860 m). Participants were asked to give an oral description of the five routes. The order of the routes was randomised and then the same order was proposed to all participants. Participants' responses and their response time for each description were recorded. These tasks were used to evaluate route mental representation ability. In fact, to solve them correctly participants had to use linear organization and left-right discrimination. These properties are considered good predictors of route mental representation because they reveal the ability to use egocentric coordinates (e.g, Prestopnik & Roskos-Ewoldsen, 2000).

Survey Tasks

1. *Landmarks Localisation* (e.g., Giusberti et al., 2009). Participants were asked to locate 10 landmarks of Bologna on an outline map printed on an A4 sheet of paper (Figure 3). The map was black and white, two-dimensional scaled with a 1:100 ratio with the environment. The landmarks were selected on the basis of a pre-test in which 10 participants had to list the most famous landmarks of Bologna. We selected those most cited: Due Torri, Piazza Maggiore, Piazza 8 Agosto, Piazza Santo Sfefano, San Domenico, San Pietro, the Railway Station, Piazza delle Mercanzie, Piazza Verdi and Via Indipendenza. Landmark names were printed in alphabetical order on an A4 sheet of paper. Participants were allowed to use any order they wished to locate them on the outline map. The experimenter recorded the order if it was different from the alphabetical one; response time was also recorded.

2. *Rotation Task* (modified by Iachini,et al., 2009). Nine landmarks in the area of Bologna were selected as stimuli based on the following criteria: they could be easily circumnavigated by moving through the environment and they were well known landmarks of Bologna. The nine buildings were combined so as to constitute the following three triads (figure 4): triad 1 = Due Torri - Piazza delle Mercanzie – Piazza Santo Stefano; triad 2 = San Petronio – Palazzo Re Enzo – Nettuno; triad 3 = Railway Station – Parco della Montagnola – Piazza 8 Agosto. Thirty maps were obtained, 10 per each triad. The triads, printed on an A4 sheet of paper, were black and white two-

dimensional scaled maps with a 1:1000 ratio with the environment. As in Iachini and co-workers (2009) study, each triad was shown in its real orientation (0°) or could be rotated by 45°, 90°, 135° or 180°. Furthermore, each triad was presented either by preserving the real positions among the buildings (five correct trials) or by altering their positions (five incorrect trials).
In order to counterbalance the different trials, we use the Latin Square design (Grant, 1948). The task was to determine whether the printed triad correctly represented the real spatial positions among the buildings. Correct judgements were scored 1 and incorrect judgements 0, maximum score = 30 and minimum = 0. The experimenter used a stopwatch to measure response latency from when participants saw the map until they responded.

These tasks measured survey representation. In fact, to solve them correctly participants rely exclusively on an abstract, internal representation characterized by a bird's-eye viewpoint of an object-centred reference system (e.g., Lawton, 1994, 1996).

The order of the nine cognitive style tasks was randomized. For each spatial task, except for the Rotation Task, the different trials were randomized as specified above then the same order was used for all participants (e.g., Roskos-Ewoldsen, McNamara, Shelton, & Carr, 1998; Nori & Giusberti, 2003).

For each task a hand-held stopwatch was used to record response time.

Procedure

On a separate day before the experimental session began, the experimenter met the participants in small groups (5/6 people) in a room at the Department of Psychology and explained the outline of the experiment to them. Then, the participants filled in the Familiarity and Spatial Cognitive Style Questionnaire. In the afternoon, as soon as the experimenter had finished checking their questionnaire answers, participants individually performed the Photo Task and the Road-Map Test. Afterwards they took part in the experimental session individually and completed the 6 spatial tasks. The experimental session lasted approximately one hour.

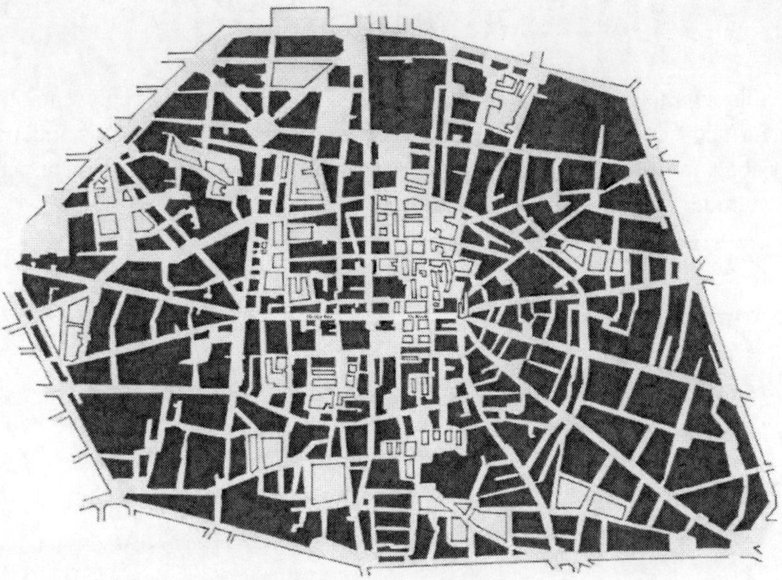

Figure 3. The outline map used by participants to locate the 10 landmarks.

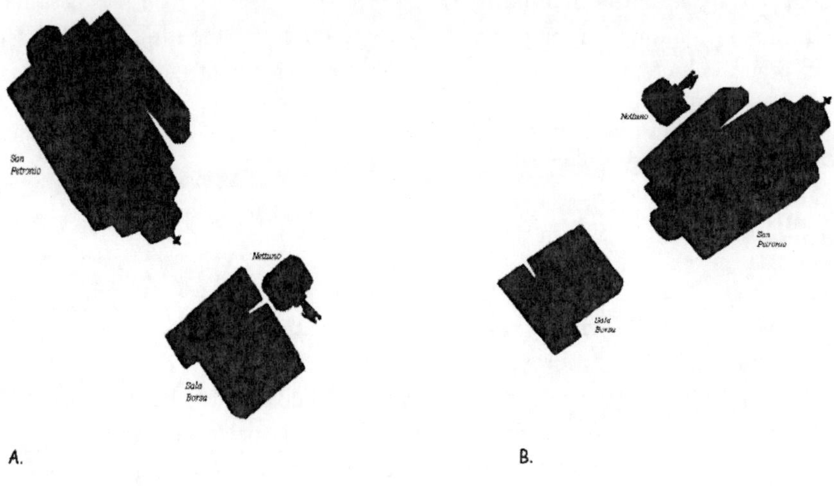

Figure 4. Examples of a correct (A.) and incorrect (B.) triad showed in the Rotation Task.

RESULTS

To evaluate the factors that predict spatial mental representation, we performed separate multiple regression analyses considering as independent variables familiarity, spatial cognitive style and sex and as dependent variables the number of hits and response times in each task.

Landmark Tasks

Landmark Production Task

The multiple regression performed on the number of hits in the *Landmark Production Task* revealed no significant differences (*Adjusted* $R^2 = .04$; $F_{[3, 36]} = 1.62$, p = n.s.), neither did the analysis performed on response time (*Adjusted* $R^2 = .04$; $F_{[3, 36]} = .62$, p = n.s.).

Landmark Recognition Task

The multiple regression performed on the number of hits in the *Landmark Recognition Task* showed a significant difference (*Adjusted* $R^2 = .35$; $F_{[3, 36]} = 8.03$, $p < .001$), that is, familiarity ($\beta = .58$; $t = 4.48$, $p < .001$) was associated with mental landmark representation. On the contrary, the multiple regression analysis conducted on response time was not significant (*Adjusted* $R^2 = .08$; $F_{[3, 36]} = 2.19$, p = n.s.).

Route Tasks

Path Task

The multiple regression performed on the number of hits in the *Path Task* showed a significant difference (*Adjusted* $R^2 = .28$; $F_{[3, 36]} = 6.06$, $p < .01$), that is, both familiarity ($\beta = .47$; $t = 3.51$, $p < .01$) and spatial cognitive style ($\beta = .29$; $t = 2.16$, $p < .05$) were predictors of high performance on the task. Nevertheless, the multiple regression analysis conducted on response time was not significant (*Adjusted* $R^2 = .08$; $F_{[3, 36]} = 2.19$, p = n.s.).

Path Description

Regarding Path Description, we considered the total score obtained by each participant on the 5 paths and performed separate multiple regression analyses on three different dependent variables: 1) quality of the description (i.e., a score of 1 was assigned if the given description allowed a person to reach the goal without getting lost in the environment and a score of zero in the opposite case); 2) richness of the description (i.e., number of landmarks reported; *range of landmarks*: 1-21); 3) ability to use spatial directional elements in verbal description (such as, to turn *right,* to turn *left,* to go *straight ahead*). Table 1 presents an example of a skeletal description of Route 1 (from Piazza del Nettuno to Railway Station) given by two participants with different levels of familiarity with Bologna.

The multiple regression performed on the quality of the description revealed a significant difference (*Adjusted* $R^2 = .29$; $F_{[3, 36]} = 6.38$, $p < .05$), that is, familiarity ($\beta = .53$; $t = 3.93$, $p < .001$) was statistically significant. The analysis performed on richness of the description also showed a significant difference (*Adjusted* $R^2 = .26$; $F_{[3, 36]} = 5.79$, $p < .01$), that is, familiarity ($\beta = .49$; $t = 3.43$, $p < .01$) was statistically significant.

The multiple regression conducted on the ability to use spatial directional elements in verbal description showed a significant difference (*Adjusted* $R^2 = .14$; $F_{[3, 36]} = 3.11$, $p < .05$), that is, both familiarity ($\beta = .30$; $t = 2.02$, $p < .05$) and spatial cognitive style ($\beta = .30$; $t = 2.04$, $p < .05$) were statistically significant. Finally, multiple regression analysis conducted on response time was not significant (*Adjusted* $R^2 = .06$; $F_{[3, 36]} = 1.96$, $p =$ n.s.).

Survey tasks.

Landmarks Localisation

Regarding Landmarks Localisation, we computed the absolute distance in millimetres (mm) on the map (scale 1:100) between the real position and the participant's response (i.e., the greater the distance, the larger the error; *range of absolute distance*: from 9300 to 90450 mm).

The multiple regression performed on the absolute distance showed a significant difference (*Adjusted* $R^2 = .11$; $F_{[3, 36]} = 2.67$, $p < .05$), that is, familiarity ($\beta = -.42$; $t = -2.80$, $p < .01$) was a predictor of participants' performance. Finally, the analysis conducted on response time revealed no differences (*Adjusted* $R^2 = .01$; $F_{[3, 36]} = 1.20$, $p =$ n.s.).

Table 1. Examples of Route 1 descriptions made by participants with different levels of familiarity with Bologna

Participant with 280 months of familiarity with Bologna	Participant with 23 months of familiarity with Bologna
You are in Piazza del Nettuno and you have to cross it. Then, you arrive at the crossing between Via Rizzoli and Via Ugo Bassi. You go straight on Via Indipendenza and pass through Altero and Piazza 8 Agosto. You go straight on Via Indipendenza and you arrive at Montagnola. You cross the road and turn left. Then, you cross the road again and you are at the railway station.	You are in Piazza del Nettuno. You go to Via Indipendenza and then at the end of the street you turn left and you are at the railway station.

Rotation Task

For the Rotation Task, we performed separate multiple regressions on each angle of rotation (0°-45°-90°-135°-180°) considering as dependent variable the number of hits and the response time. At 0°-45°, the regression model showed no significant differences (0°: *Adjusted* $R^2 = .04$; $F_{[3, 36]} = .58$, p = n.s.; 45°: *Adjusted* $R^2 = .02$; $F_{[3, 36]} = .31$, p = n.s.). At 90°, the analysis showed a significant difference (*Adjusted* $R^2 = .15$; $F_{[3, 36]} = 3.45$, $p < .05$), that is, spatial cognitive style ($\beta = .37$; $t = 2.54$, $p < .05$) was a predictor of participants' performance. At 135° and 180°, the analysis revealed a significant difference (135°: *Adjusted* $R^2 = .24$; $F_{[3, 36]} = 5.19$, $p < .05$; 180°: *Adjusted* $R^2 = .18$; $F_{[3, 36]} = 4.02$, $p < .05$). In both cases, the predictive factors were familiarity (135°: $\beta = .35$; $t = 2.53$, $p < .05$; 180°: $\beta = .30$; $t = 2.13$, $p < .05$) and spatial cognitive style (135°: $\beta = .38$; $t = 2.79$, $p < .05$; 180°: $\beta = .34$; $t = 2.40$, $p < .05$). Response times were never significant, regardless of the angle of rotation (0°: *Adjusted* $R^2 = .07$; $F_{[3, 36]} = .96$, p = n.s.; 45°: *Adjusted* $R^2 = .03$; $F_{[3, 36]} = .44$, p = n.s.; 90°: *Adjusted* $R^2 = .05$; $F_{[3, 36]} = 1.78$, p = n.s.; 135°: *Adjusted* $R^2 = .04$; $F_{[3, 36]} = .60$, p = n.s.; 180°: *Adjusted* $R^2 = .03$; $F_{[3, 36]} = .45$, p = n.s.).

Finally, we found no effects linked to gender differences in any of the tasks.

CONCLUSION

In this study, we set out to investigate which model of spatial representation acquisition fits best with participants' performance in a real environment learned by means of "free-exploration" and considering months of stay as level of familiarity with the town. Our results showed that for landmark representation, an important distinction has to be made with respect to tasks. In fact, the first task (Landmark Production), in which participants had to list ten landmarks of Bologna, assessed only semantic knowledge of the town. Instead, the second task (Landmark Recognition) measured figurative memory and the ability to discriminate a town landmark among similar fillers. Navigation through the environment was not required in the first task, because this type of knowledge is encyclopaedic in nature. In fact, Landmark Recognition does not require spatial ability. However, performance of the second task required having explored the environment previously. For this reason, Siegel and White (1975) considered it the first step in spatial representation and in Pazzaglia and co-workers' model (2000) a landmark style corresponds to a poor and less efficient style of navigation. Differently, Montello's model (1998) suggests that accuracy of detail increases with familiarity. Our results showed that, regardless of familiarity and cognitive style, in Landmark Production participants were able to list town landmarks and that in Landmark Recognition familiarity predicted performance. These data seem to support both Siegel and White's model and that of Pazzaglia and co-workers. Regarding Montello's framework, familiarity should predict performance on both tasks. However, the list length constraint may have prevented participants with higher town familiarity from making longer and more detailed lists.

In route representation, we found that familiarity predicted participants' performance only in the ability to correctly reassemble a path, in the richness and in the ability to use spatial elements in verbal description spatial, while spatial cognitive style is predictive of a good performance only regarding the ability to reassemble a path correctly and regarding the richness and the ability to use spatial elements in verbal description. These results support both Siegel and White's and Pazzaglia et al's models but not Montello's prediction. In fact, performances improved with familiarity and, as suggested by Pazzaglia and co-workers (2000), the higher spatial ability is the higher the performances obtained, because survey cognitive style individuals can easily shift strategy depending on the task requirement, whereas route cognitive style individuals use their preferred strategy and landmark cognitive style individuals are the only ones unable to use spatial references to solve the tasks. These results are also in line with those of Iaria and co-workers (2003) who found that human participants spontaneously adopt different strategies to solve a navigation task and that these strategies lead to differential neural activities. Their findings provide evidence of a shift in the neural mechanisms of the human brain and are consistent with and extend previous work on rodents (Packard & McGaugh, 1996; White & McDonald, 2002).

In survey tasks, high familiarity always predicts good performance, but spatial cognitive style is predictive only when the task requires mental rotation, specifically at 90°, 135° and 180°. This type of estimation involves allocentric frames of reference that improve with increased familiarity and with high spatial abilities. Therefore, our evidence agrees with Siegel and White's and Pazzaglia and co-workers' models. However, based on the predictions of Montello's model metric and nonmetric knowledge are not acquired simultaneously, at least in a real environment. It is important to note that, differently from research in line with this model (e.g., Klatzky, et al., 1990; Loomis, et al., 1993), our study was carried out in a large natural complex environment, that is, a city, not in a simple college environment or in a building and not in a restricted laboratory setting. Our ecological setting allows us to propose a new model to explain individual differences in moving successfully through the environment.

In fact, we propose the Environmental Knowledge Model (EKM) in which the higher the spatial cognitive load, the more spatial cognitive style predicts success in navigation. Therefore, individuals who are able to shift strategy and to have a flexible mental representation perform better than individuals with an inflexible representation of space and low spatial competence. Regardless of their familiarity with the environment, individuals with high spatial abilities fit easily into the environment despite its complexity. Thus, individual differences come into play only when spatial ability requirements increase. In any case, environmental familiarity is crucial in each environmental mental representation (landmark, route and survey). In fact, regardless of their spatial cognitive style (see figure 5), the higher individuals' familiarity with the environment is, the better their performances are.

Gender differences are a different matter. We did not observe any differences between males and females, in line with previous studies on real-world spatial tasks that found minimal differences when environments were familiar and much greater differences when they were unfamiliar (see Montello, Lovelance, Golledge, & Self, 1999). It is also possible that we did not observe differences because our sample (although we balanced participants for degree of familiarity) was too small to show the presence of a gender effect.

In summary, we can affirm that both familiarity with the environment and spatial cognitive style are important for acquiring and representing spatial information, but their involvement depends on task demands.

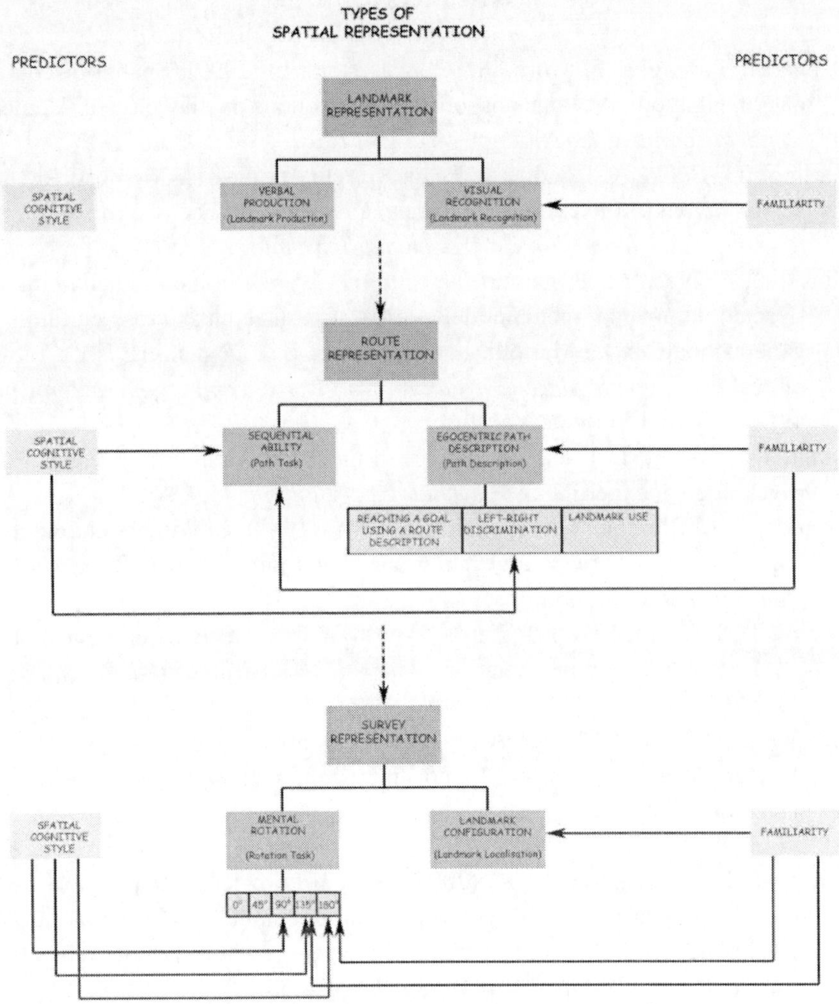

Figure 5. Environmental Knowledge Model (EKM): predictors in relation to task demands.

Acknowledgments

We thank Selena Brusa for her help in collecting the data.

References

Denis, M., Pazzaglia, F., Cornoldi, C. & Bertolo, L. (1999). Spatial discourse and navigation: An analysis of route directions in the city of Venice. *Applied Cognitive Psychology, 13,* 145-174.

Gärling, T., Böök, A., Lindberg, E. & Nilsson, T. (1981). Memory for the spatial layout of the everyday physical environment: Factors affecting rate of acquisition. *Journal of Environmental Psychology*, *1*, 263-277.

Giusberti, F., Nori, R. & Mercuri, N. (2009). Vivere e orientarsi nella città di Cesena: la mappa mentale della città, fra differenze cross-culturali e generazionali. In A. Martelli, F. Giusberti & E. Cicognani (Eds.), *Cesena oggi. Vivere, condividere e orientarsi tra genere, genti e generazioni* (pp. 49-76). Cesena: Il Ponte Vecchio.

Grant, D. A. (1948). The Latin Square principle in design and analysis of psychological experiments. *Psychological Bulletin*, *45*, 427-442.

Herman, J. F., Blomquist, S. L. & Klein, C. A. (1987). Children's and adults' cognitive maps of very large unfamiliar environments. *British Journal of Developmental Psychology*, *5*, 61-72.

Iachini, T., Ruotolo, F. & Ruggiero, G. (2009). The effects of familiarity and gender on spatial representation. *Journal of Environmental Psychology*, *29*, 227-234.

Iaria, G., Petrides, M., Dagher, A., Pike, B. & Bohbot, V. D. (2003). Cognitive strategies dependent on the hippocampus and caudate nucleus in human navigation: variability and change with practice. *The Journal of Neuroscience*, *23(13)*, 5945-5952.

Ishikawa, T. & Montello, D. R. (2006). Spatial knowledge acquisition from direct experience in the environment: Individual differences in the development of metric knowledge and the integration of separately learned places. *Cognitive Psychology*, *52*, 93-129.

Klatzky, R. L., Loomis, J. M., Golledge, R. G., Cicinelli, J. G., Doherty, S. & Pellegrino, J. W. (1990). Acquisition of route and survey knowledge in the absence of vision. *Journal of Motor Behaviour*, *22*, 19-43.

Landau, B., Spelke, E. & Gleitman, H. (1984). Spatial knowledge in a young blind child. *Cognition, 16*, 225-260.

Lawton, C. A. (1994). Gender differences in way-fi nding strategies: Relationship to spatial ability and spatial anxiety. *Sex Roles, 30,* 765-779.

Lawton, C. A. (1996). Strategies for indoor way-finding: The role of orientation. *Journal of Environmental Psychology, 16,* 137-145.

Loomis, J. M., Klatzky, R. L., Golledge, R. G., Cicinelli, J. G., Pellegrino, J. W. & Fray, P. A. (1993). Nonvisual navigation of blind and sighted: Assessment of path integration ability. *Journal of Experimental Psychology: General*, *122*, 73-91.

Miller, L. K. & Santoni, V. (1986). Sex differences in spatial abilities: Strategic and experimental correlates. *Acta Psychologica, 62*(3), 225-235.

Money, I., Alexander, D. & Walker, H. T (1965). *Manual: A stundardised road-map test of direction sense*. Baltimore: Johns Hopkins Press.

Montello, D. R. (1998). A new framework for understanding the acquisition of spatial knowledge in large-scale environments. In M. J. Egenhofer, & R. G. Golledge (Eds.), *Spatial and temporal reasoning in geographic information systems* (pp. 143-154). New York: Oxford University Press.

Montello, D. R., Lovelance, K. L., Golledge, R. G. & Self, C. M. (1999). Sex-related differences and similarities in geographic and environmental spatial abilities. *Annals of the Association of the American Geographers*, *89*, 515-534.

Nori, R. & Giusberti, F. (2003). Cognitive styles: Errors in the directional judgements. *Perception, 32,* 307-320.

Nori, R. & Giusberti, F. (2006). Predicting cognitive styles from spatial abilities. *American Journal of Psychology*, *119(1)*, 67-86.

Packard, M. G. & McGaugh, J. L. (1996) Inactivation of hippocampus or caudate nucleus with lidocaine differentially affects expression of place and response learning. *Neurobiological Learning Memory*, *65*, 65-72.

Pazzaglia, F. & De Beni, R. (2006). Are people with high and low mental rotation abilities differently susceptible to the alignment effect? *Perception*, *35(3)*, 369-383.

Pazzaglia, F. & De Beni, R. (2001). Strategies of processing spatial information in survey and landmark-centered individuals. *European Journal of Cognitive Psychology, 13(4)*, 493-508.

Pazzaglia, F., Cornoldi, C. & De Beni, R. (2000). Differenze individuali nella rappresentazione dello spazio: Presentazione di un questionario autovalutativo [Individual differences in spatial representation and in orientation ability: Presentation of a self-report questionnaire]. *Giornale Italiano di Psicologia*, *3*, 627-650.

Prestopnik, J. L. & Roskos-Ewoldsen, B. (2000). The relations among wayfinding strategy use, sense of direction, sex, familiarity, and wayfinding ability. *Journal of Environmental Psychology, 20,* 177-191.

Roskos-Ewoldsen, B., McNamara, T. P., Shelton, A. L. & Carr, W. (1998). Mental representations of large and small spatial layouts are orientation dependent. *Journal of Experimental Psychology: Learning, Memory and Cognition*, *24*, 215-226.

Salmaso, D. & Longoni, A. M. (1985). Problems in the assessment of hand preference. *Cortex, 21(4)*, 533-549.

Siegel, A. W. & White, S. H. (1975). The development of spatial representations of large-scale environments. In H. W. Reese (Ed.) *Advances in child development and behavior* (vol. *10*; 9-55). New York: Academic Press.

Vingerhoets, G., Lannoo, E. & Bauwens, S. (1996). Analysis of the Money Road-Map Test Performance in Normal and Brain-Damaged Subjects. *Archives of Clinical Neuropsychology*, *11(1)*, 1-9.

Ward, S. L., Newcombe, N. & Overton, W. F. (1986). Turn left at the church, or three miles north: A study of direction giving and sex differences. *Environment and Behavior, 18,* 192-213.

White, N. M. & McDonald, R. J. (2002). Multiple parallel memory systems in the brain of the rat. *Neurobiological Learning Memory*, *77*, 125-184.

In: Spatial Memory
Editor: Jacob B. Thomas, pp. 145-167
ISBN: 978-1-61668-139-5

Chapter 6

VISUAL PERCEPTION, OPTICAL ILLUSIONS AND FREQUENCY ANALYSIS: AN INTER-RELATED APPROACH

Mina Ray and Amitabha Basuray*

Department of Applied Optics and Photonics, University of Calcutta, 92, A.P.C. Road, Calcutta-700 009, India

INTRODUCTION

Optics is considered to be the oldest branch of science as light and vision have long been considered to be the central for understanding the world in which we live. Since, ancient day's vision is regarded as the supreme sense for the acquisition of knowledge, and light is the key to unlock the secrets of the universe. The mechanism of vision is regarded as the formation of an image of the object we see, by the lens of the eye, on to a light sensitive-membrane in the back of the eye, called the retina. The retina serves as a transducer and converts the image pattern formed by the light into neuronal signals. These signals are processed in different parts of the brain and we finally see. However, the major problem in visual perception is that what people see is not simply a translation of the image on the retina. Thus people interested in perception have long struggled to explain what visual processing does to

* Corresponding author: E-mail: abasuray@rediffmail.com

create what we actually see. Visual perception is the ability to interpret information from visible light reaching the eye. Hermann von Helmholtz [1] the pioneering man in the study of visual perception in modern times, felt that the poor quality of information gathered by the eye can not make vision possible and concluded that vision could only be the result of some form of unconscious inferences that is, to make assumptions and conclusions from incomplete data, based on inference data. Gestalt psychologists [2] have guided the study of how people perceive visual components as organized patterns instead of many different parts. Then onwards, the visual perception is considered a combination of physiological and psychological processes and different groups are thinking of many models for visual perception. During the last three decades many computational models of vision were developed which explained visual phenomena relatively closer to real life situation.

In the visual cognition process, the information received is analyzed into its different attributing components, which subsequently synthesized to reach into a complete perceptual experience. Though there have been remarkable advancements, in recent years, in the understanding of visual process and its underlying mechanisms still there is enormous mystery surrounding the question of how we see. When light enters the eye, it first passes through the cornea, then the aqueous humor, lens and vitreous humor. Ultimately it reaches the retina, which is the light-sensing structure of the eye. The retina contains two types of cells, called rods and cones. Rods handle vision in low light, and cones handle color vision and detail. Much of the information that we possess of the anatomical aspects of the visual systems, are based on observation on animals[3]. By using tissues and enucleated eyes from human cadavers, visual scientists have been able to examine the gross morphology of the human visual system though the knowledge about the active neurophysiological results is rare. However, the knowledge of the primitive nature of eye can hardly justify the overall capability of eye in the visual cognition process. Research in the area of visually enabled cognitive systems focuses on various theories, at different levels of abstraction, regarding requirements, architectures, forms of representation, kinds of ontology and knowledge, and varieties of mechanisms relevant to integration and control of vision systems. This requires a vast effort in a multidisciplinary understanding of cognitive processes, involving studies in cognitive psychology, and neuroscience [4, 5].

The visual scientists are coming to a consensus that three semi-independent channels namely, physical, neurophysiological and neuropsychological processing systems are responsible for the cognition process and the processed information is then sent to the brain for the cognition. This further forces one to think of the logical necessity for some kind of recombination process somewhere or at some time in the central visual nervous system. Neuropsychological processing is semi-independent as it depends on the training of the eye, that is, on experience, but neurophysiological processing has some objectivity and is dependent on the nature of the physical processing. In the study of visual systems, visual illusions, that is, the cases when the inference process goes wrong, have played an important role in understanding and deciding what sorts of models are to be considered. In the present article, we are interested to discuss the optical illusions from the angle that may be explicable in neurons-physiological terms, that is behind such illusions there are always some significant objective elements. There are some differences of opinion even in case of formation of pattern on the retina. Often asked question is whether the image of the object or a pattern signifying the spatial frequency of the object is formed on the retina, which renders the information to our brain for final processing [6,7]. In the second case some kind of frequency filtering may be incorporated in the processing of image at the brain, which in turn guides the visual process as it is found that both psycho-physics and neuro-physiology play important roles in cognitive science. We will consider here some important visual illusions and discuss them on the basis of objective elements of patterns formed at the retina and their probable processing at the brain.

Human Visual System

As already mentioned that in the visual process the eye lens produces an optical energy distribution on the retina and then in the second stage a spatial processing takes place, which is essentially neural. The third stage is that of the final perception of the picture, which is psychophysical in nature and therefore beyond our scope of discussion here. To understand the relationship between original picture and the perceived picture a model is to be framed to define the relationship underneath as the perceived picture can neither be directly measured nor be described by some mathematical function. The most important aspect of the visual process is that a huge amount of data is processed in visual cognition. As the processing is neural for which switching speed is of the order of millisecond and persistence of vision of the order of one twentieth of a second, the frame per second is only few. It is true that unlike digital processing in computer the processing here is analog in nature as well as data is processed in parallel format; but that does not explain fully the nature of processing. Many vision scientists feel that the human eye can be considered to be a biological adaptive optics (AO) system. Adaptive optics is a technique that can correct for the eye's aberrations and provides diffraction-limited resolution. An AO system is essentially one in which the optical element is adapted to correct for the image blur present at that point [8]. Such a system consists of three main elements: the sensor, the corrector and the controller. This model can explain the higher resolution achieved by the poor quality eye lens but not the other aspects of visual system.

One of the most popular theoretical models of the visual system is based on the idea that a system of spatial frequency filters exists in the visual system. The basic idea embodied in what has now come to be called the "Fourier Model" [6,7,9] is a classic example of the multiple-module type of theory in that it postulates the existence of a system of two-dimensional "filters" or feature detectors, sensitive to the individual spatial frequency components of a visual stimulus. This model hypothesizes that the visual system has a number of channels or modules that independently respond to these spatial frequencies in a way that analyses the two-dimensional retinal image into the responses of these filters. Responses of the visual systems of cat and monkey to sinusoidal patterns have been measured at the ganglion cells and cells were found to be tuned to limited bands of spatial frequency. It was also found that the cortical neurons are sensitive to different orientation of the grating [10, 12]. This suggests that the visual system of humans is similarly organized in channels that are sensitive to the frequency and orientation of the Fourier components of the spatial patterns. Based on these studies several theories state that visual information is processed in two different streams in the primate brain – the dorsal and the ventral streams. The function of the dorsal stream is the analysis of the spatial relations between objects and the function of the ventral is the recognition of the objects [13]. Goodale and Milner [14] re-interpreted the theory and suggested that the function of the dorsal stream is not to analyze the location of the objects but rather to guide the manipulation of the objects; whereas, the function of the ventral stream is to perform computations that are necessary for object recognition and conscious perception. Several other models with some modifications are also proposed. However, all these models are suitable for partial solution of the involved visual cognition process. Nevertheless, all these studies are very important for understanding of vision.

Image Processing

Duffieux in 1946 [15] first established that a lens system produces at the back focal plane the Fourier transform of the distribution at the front focal plane. Then onwards an image is in general described not only in terms of image intensity distribution, but also by its spectrum in the form of a two-dimensional Fourier transforms. It is seen that a lens system transfers different spatial frequency components with different attenuation and behaves as a low pass filter [16]. The spatial frequency is defined in terms of infinite lines per mm, however further studies reveal that the optical transfer function (OTF) depends significantly on size and shape [17] of the object structure for which the spatial frequency is defined. Fourier transform has now become an indispensable tool for the description of optical system. The spectral approach has played a major role in image processing theory, leading to significant technical developments in pattern recognition. The application of Fourier methods in physical optics also reduces the processing time and optical physicists and technicians are of the opinion that frequency plane processing is superior to image plane processing, The vision scientists also initiated spectrum plane processing for pattern recognition practically during the same period – this is not for the eye as an image forming system [18,19]. This was an important advance, since it provided a united approach for describing the flow of spatial data in a cascade of optical and neural systems. Some discovery of psychophysical and neuro-physical evidence [20,21] indicates the existence of neural channels, which are selectively sensitive to frequency of the viewed spatial pattern. This suggests that the visual system may perform the operation of two-dimensional Fourier transform.

At this point the question arises that, whether the image pattern or the spatial frequency spectrum pattern is formed on the retina. As in the case of eye lens the object distance is always much higher than the focal length and the image is formed in a region very close to the focus while on the other hand, at the focus we get the spatial spectrum of the object. The spectrum plane and the image plane coincide when the image is at the infinity, whereas the separation between the focal plane and the image plane is the maximum when the object is at the least distance of distinct vision. As the surface or the image plane is fixed the lens focal length is reduced to form the image on the retina. The change to be introduced in the focal length is of the order of a fraction of a millimeter. We may consider that the information is processed in two steps – the image is formed at the retina and the image is recorded intermittently and secondly the in between time is used to detect the spatial spectrum through Maxwellian view. In the Maxwellian view the Fourier transform of the entrance pupil is recorded on the retina and one views the Fourier transform of the spectrum that is the image formed at the exit pupil. This spectral pattern recording is used for processing of the image. The spectrum formed at the retina may be processed by insertion of some hypothetical aperture in the frequency plane or by the nature of the visual cortex [22].

Optical Illusion

An optical illusion is characterized by visually perceived images that differ from objective reality. The information gathered by the eye is processed in the brain to give a perception that does not tally with a physical measurement of the stimulus source. It is a widely followed method used by the vision scientists to analyze the models with these illusionary effects to evaluate their performance. It is herein proposed that the present model of visual perception may shed a consistent light on many of the known paradoxes namely illusions. We present here some objects and their Fourier spectrum [23] to understand the basic logical frame of the present analysis.

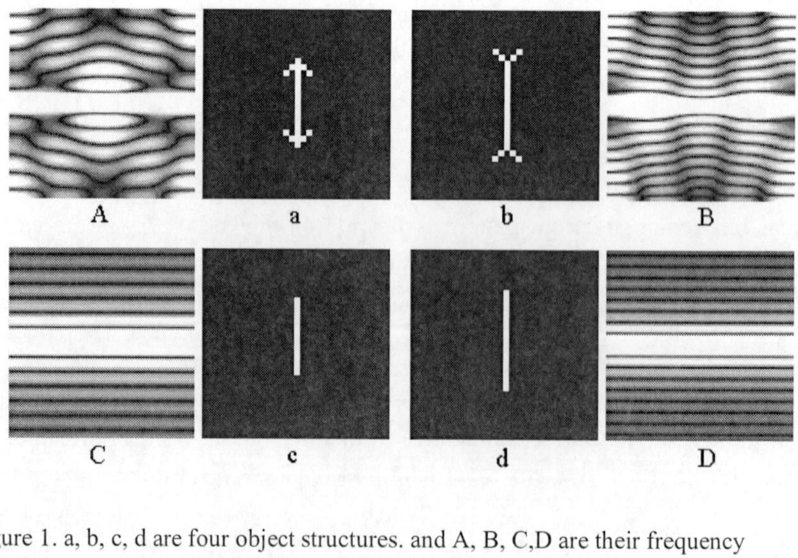

Figure 1. a, b, c, d are four object structures. and A, B, C,D are their frequency spectrum respectively. (These figures are taken from Reference [23])

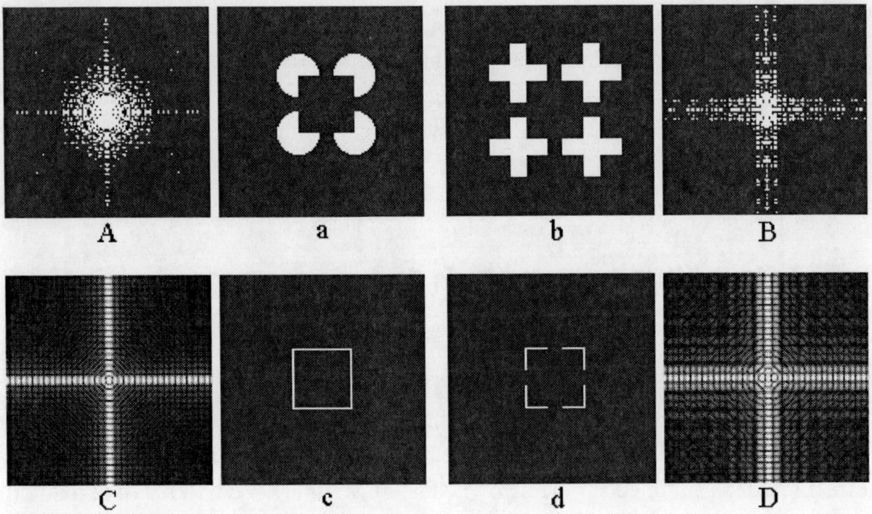

Figure 2. a, b, c, d are the object structures and A, B, C, D are their Fourier spectrum respectively. (These figures are taken from Reference [23])

Of these figures Figure 1(a,b) is the famous Muller-Lyer illusion consisting of the perception of the shaft of the arrow in Figure 1(a) as shorter than the shaft of the arrow in Figure 1(b) though they are equal. Figure 1(A) and 1(B) are their spectra. Figure 1(c) and 1(d) are two lines having unequal lengths and 1(C) and 1(D) are their respective Fourier Transforms. Figure 2(a) shows appearance of a black square framed by four "pacmen" shapes and 2(A) is its Fourier spectrum. Figures 2c and 2d are two squares: one with close contour and the other with open contours, and figure 2C and 2D are their spectrum respectively. Here also some dark cross like structure appears in the case of open contours square. There have been many attempts to explain these illusions from different angles [24,25]. We will see here how the combination of image and spectrum analysis can explain the formation of such illusions. It is apparent that each pattern is associated with a particular distribution in the spectrum plane, which might be called the signature of the pattern. The image formed is again given by the inverse Fourier transform of the spectrum pattern. However lens system behaves like a low pass filter therefore all the frequencies do not contribute to the formation of image; moreover, different frequencies are attenuated to different extent. Thus the image is never an exact replica of the object pattern. In the next section we will discuss how to apply spatial filtering technique in analysis of image on the retina as in image processing and pattern recognition. At this point it is to be noted that if we introduce some modification in the spectrum corresponding to some object structure, the image distribution will be modified accordingly and this is the technique that is being followed in widely used image processing using spatial filtering. With this background we will discuss in details the Muller-Lyer illusion.

Muller-Lyer Structure for Optical Illusion

The object structure chosen is that used to illustrate Muller-Lyer illusion, that is, lines of equal lengths as shown in (Figure 3)

It is apparent from the figure 3(b) that the effect is same even if shaft joining the arrowhead is missing. From Figure 3(c, d) it is apparent that the illusionary effect is less pronounced if the arrowheads are smaller. We will try to explain all these effects from the suggested model of image processing. The illusion is subjective in nature and is due to the transformations to which the visual image is subjected and these transformations depend on the visual processing system. Our intension is to show that some kind of spatial filtering also causes similar illusion effects. The mathematical formulation of the problem is also very simple as we may consider that the object structure comprises five rectangular slit functions: one, the shaft of the double arrow and four others inclined to the axis forming the arrowheads. The spectrum plane distribution can be obtained by the superposition of the distribution pattern due to each slit distribution pattern of all the slits as the Fourier transform is a linear transform. At the time of addition it is to be noted that when the slit is shifted from the center as in the case of slits of arrowhead, an additional phase factor is involved and some co-ordinate transformation is necessary [26]. However, many software routines are also available for obtaining two dimensional Fourier transformations; we have used here Matlab software.

Computational Results

We first start with the pattern depicted in figure 3a.and make the Fourier transform of the pattern, the 3D plot of the power spectral distributions are depicted in figure 4(a) and 4(b) respectively where F_{IN} stands for the Fourier transform of the inward arrow and F_{OUT} for the outward arrow.

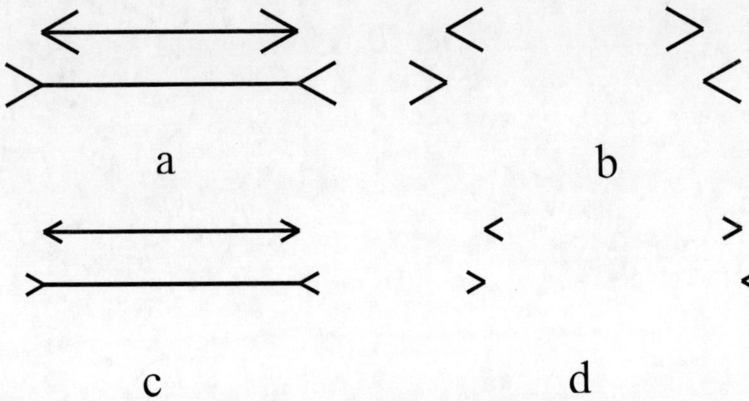

Figure 3. (a) Illusion problem of lines of equal lengths (tailed lines) appearing to be of unequal length and (b) same problem with central joining line removed (c) same as (a) with smaller arrow head and (d) same as (b) with a smaller arrow head.

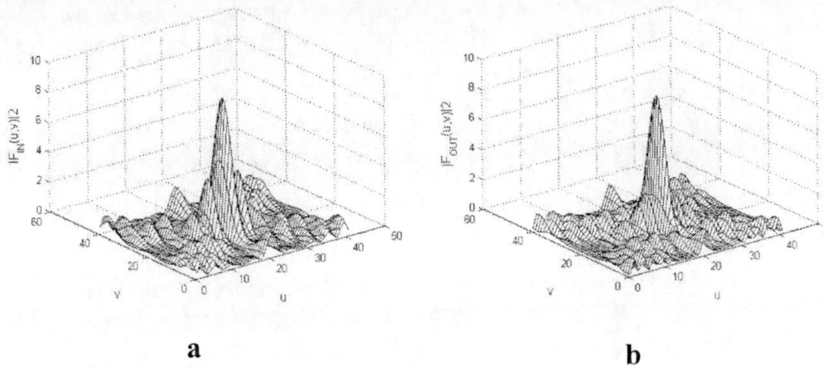

Figure 4. 3D-plot of power spectral distribution (a) $|F_{IN}(u,v)|^2$ and (b) $|F_{OUT}(u,v)|^2$ (u,v) are in the range (-2 to +2).

The image distribution is obtained by evaluating the inverse Fourier transform of this spectrum. The inverse Fourier transform of $F_{IN}(u,v)$ and $F_{OUT}(u,v)$ distributions give the images I_{IN} and I_{OUT} as shown in Figure 5(a) and 5(b) respectively. contour plots are given in Figure 6(a) and 6(b) respectively.

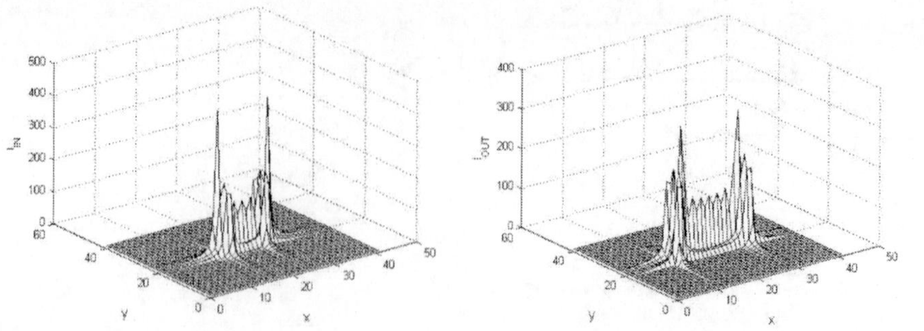

a **b**

Figure 5. 3D plot of intensity distribution (a) I_{IN} and (b) I_{OUT} corresponding to inward and outward arrows respectively.

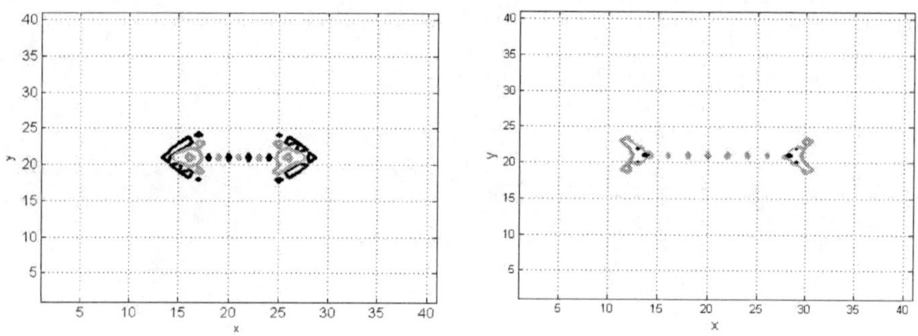

a **b**

Figure 6. contour plot of the image of (a) inward (b) outward arrows.

It is evident from the figure 5(a) & 5(b) and 6(a) & 6(b) that the effective lengths are different in the image plane and this gives an explanation of the Muller-Lyer illusion. It should be noted here that there is some implied truncation in the transform plane. This effect will be more prominent if a directional selection of the spectrum plane distribution is made. The figure 7(a) and 7(b) are the distributions along the u-axis, that is, along (u, 0) line of the frequency patterns.

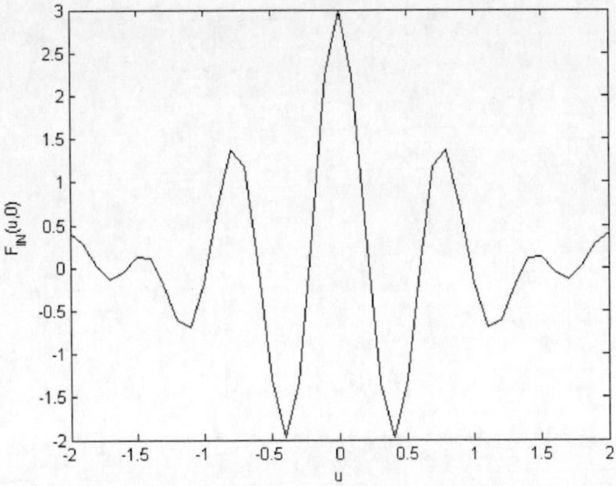

(a)

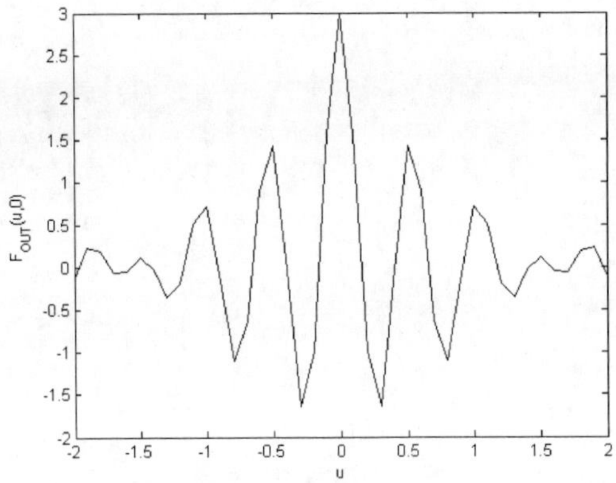

(b)

Figure 7. Axial (u, 0) distribution of the spectral pattern (a) inward and (b) outward arrowheads.

The Fourier transform of these patterns will give the axial intensity distributions along (x, 0).

It is quite evident that the image of the shaft of the outward arrow is longer. Next we will repeat computation of all these results for the arrowheads of figure 1(c) with shorter arrowheads but shaft size remaining the same. It is apparent that the illusion is still there but less pronounced. In Figure 9(a,b) and 10(a,b) are the figures depicting the respective axial distributions of spectrum and intensity of two arrowheads. The results confirmed our observation, the image formed on the retina shows that the shaft with outward arrowheads is longer but elongation is less in comparison to the earlier case.

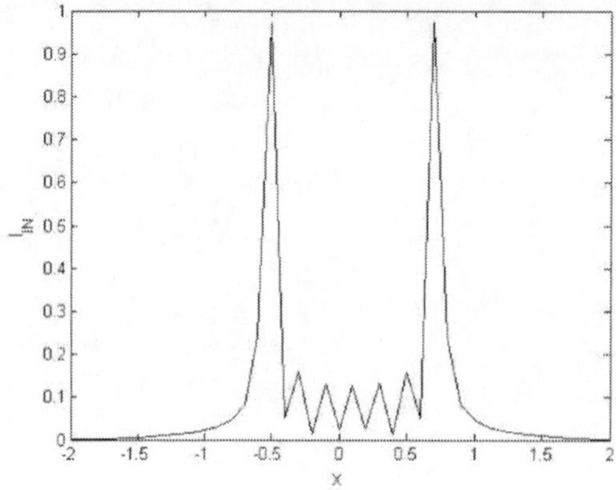

(a)

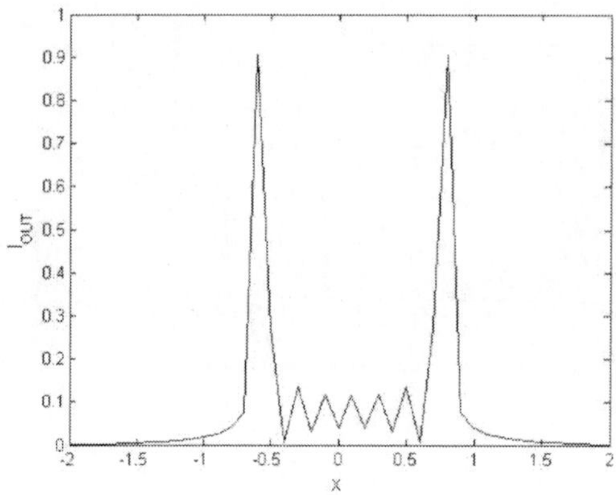

(b)

Figure 8. Image intensity distribution (a) $I_{IN}(x,0)$ and (b) $I_{OUT}(x,0)$ for inward arrow and outward arrow respectively.

Arrowheads without Shaft

It is seen from the figure 1(b) that the illusion persists even if the shaft lines are replaced by gaps indicating that the illusion is due to the position and dimension of the arrow-heads and not due to the line joining them. The illusion effects may be explained by the similar logic as in the previous cases. The figures 11 (a and b) are the spectral power density distributions at the back focal plane for the inward arrowhead and outward arrowheads without the shafts.

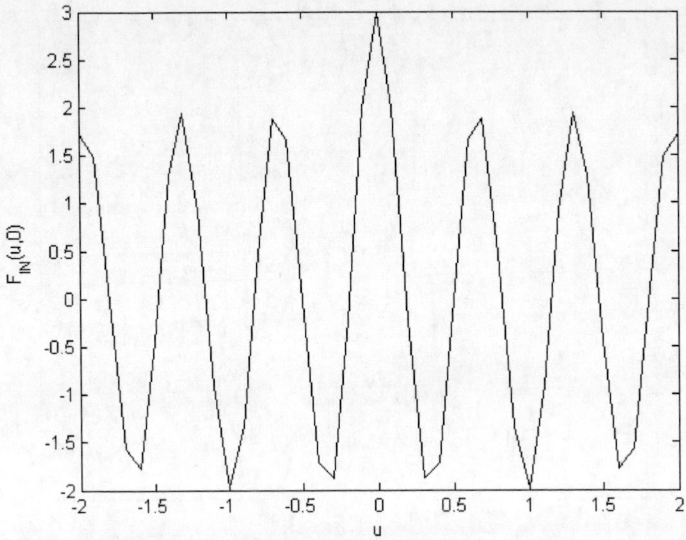

(a)

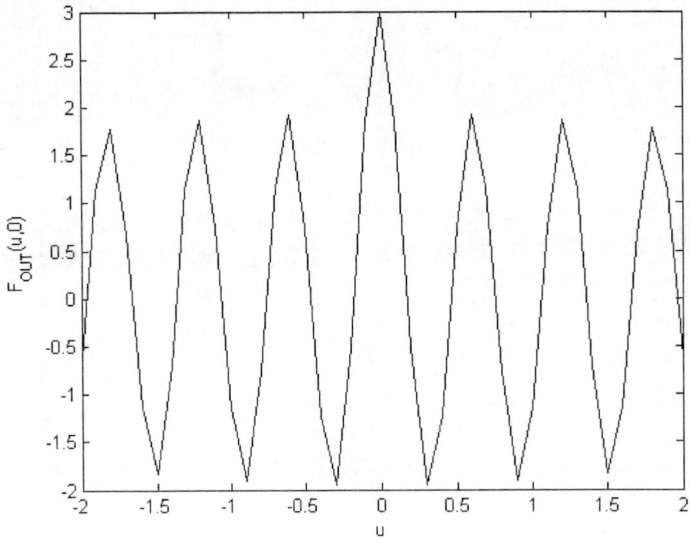

(b)

Figure 9. Axial (u, 0) distribution of the spectral pattern for (a) inward and (b) outward arrowheads with reduced arrowhead size, shaft size remaining the same.

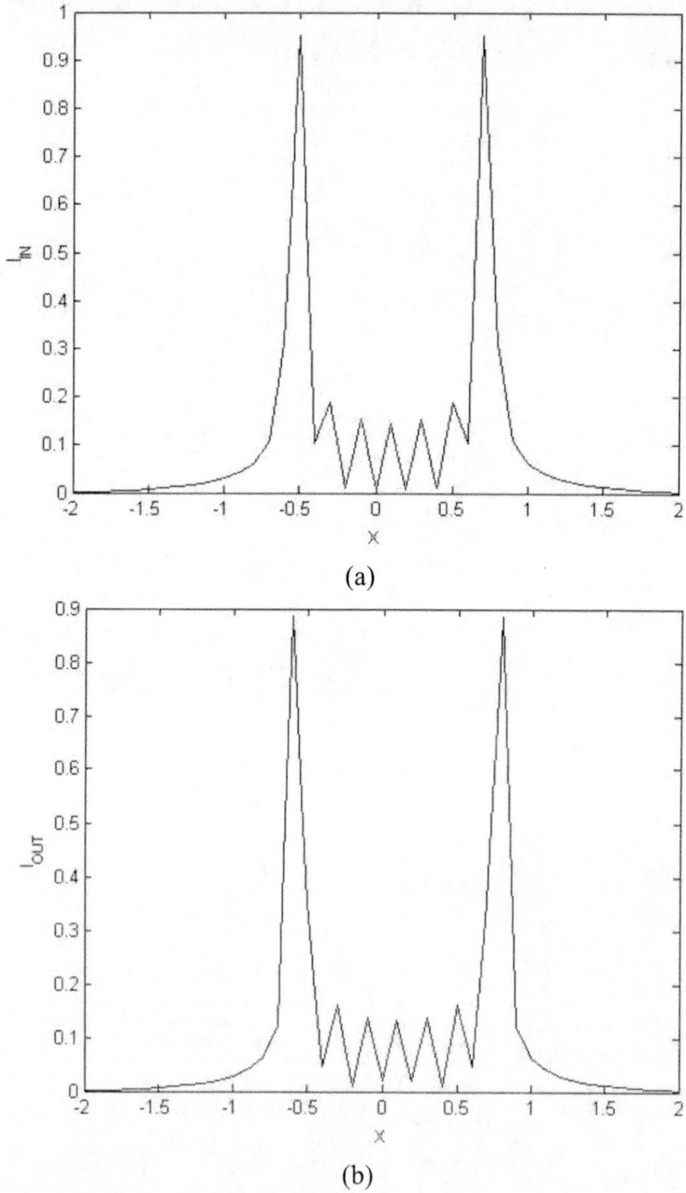

(a)

(b)

Figure 10. Image intensity distribution (a) $I_{IN}(x,0)$ and (b) $I_{OUT}(x,0)$ for inward arrow and outward arrow respectively.

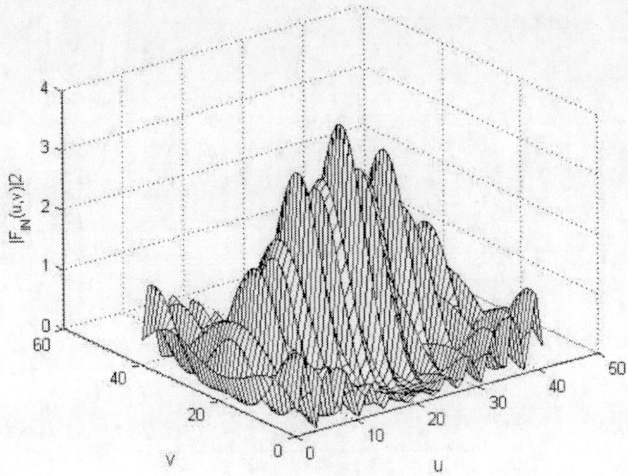

(a)

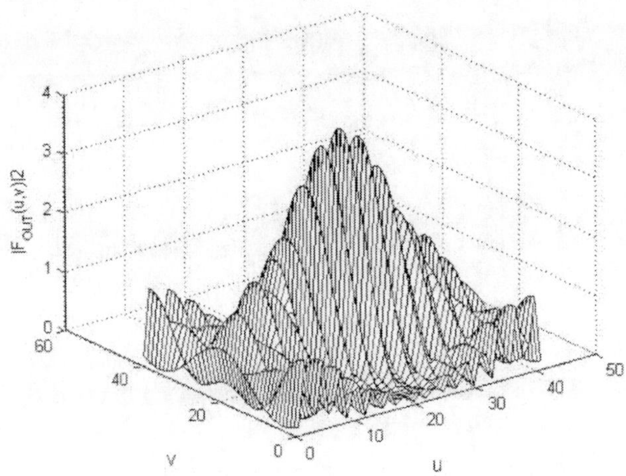

(b)

Figure 11. 3D-plot of power spectral distribution(a) $|F_{IN}(u,v)|^2$ and (b) $|F_{OUT}(u,v)|^2$ (u,v) are in the range (-2 to +2).

Figure 12 (a) and (b) are the contour plots of the power spectrum distribution, The nature of fringes resemble the fringe pattern that would be observed in the optical Fourier transform plane. It is seen from these contour diagram that the separations between the lobes are different in two cases – greater in case of inward arrow and less in the outward arrow. These separations play the role in illusion.

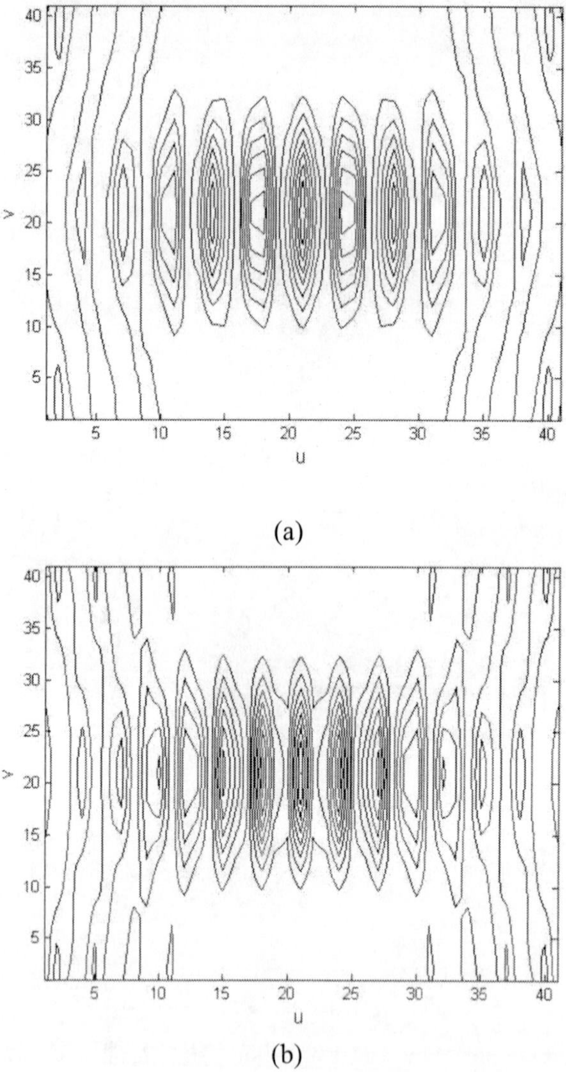

Figure 12. Contour diagram of power spectral distribution of (a) inward arrow and (b) outward arrow without shaft.

Taking the inverse Fourier transform of F_{IN} (u,v) and F_{OUT}(u,v) the intensity distributions at the image plane are obtained and the contour of these functions gives the nature of the images as shown in figure 13(a and b) and 14(a and b) respectively.

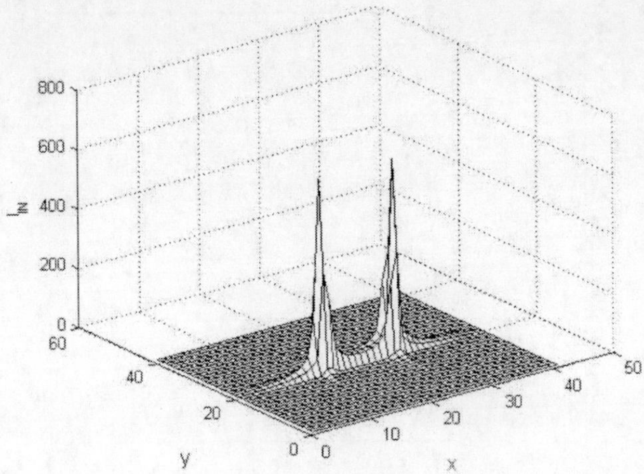

(a)

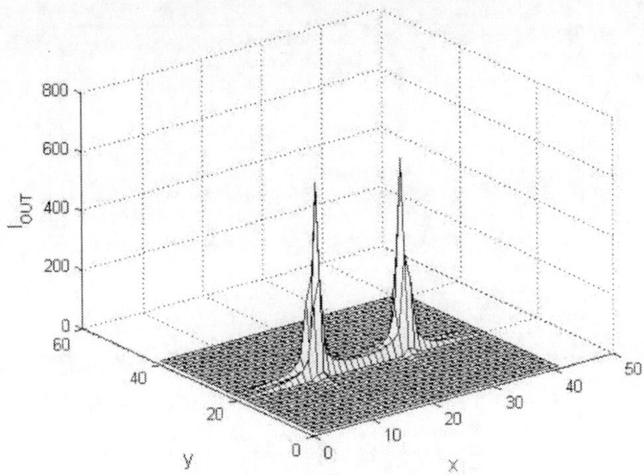

(b)

Figure 13. Image intensity distribution (a) $I_{IN}(x,0)$ and (b) $I_{OUT}(x,0)$ for inward arrow and outward Arrow without shaft respectively.

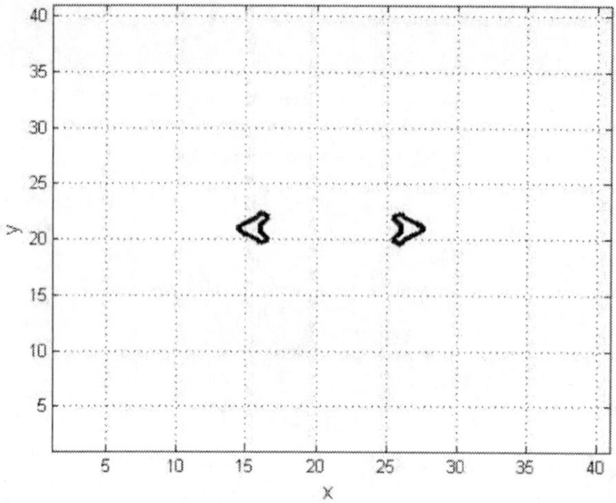

(a)

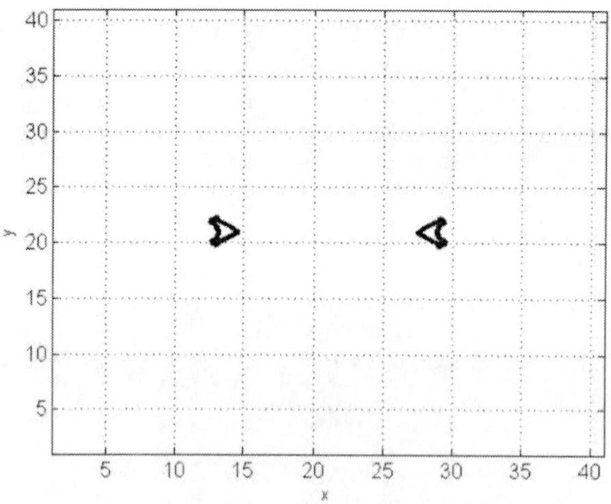

(b)

Figure 14. contour plot of the image of (a) inward (b) outward arrows without shaft.

Figures 15(a,b) and 16(a and b) are the plot of spectral patterns along (u,0) and their respective inverse Fourier transforms. These confirm the reason of illusion. Thus we have explained all the aspects of Muller-Lyer illusion.

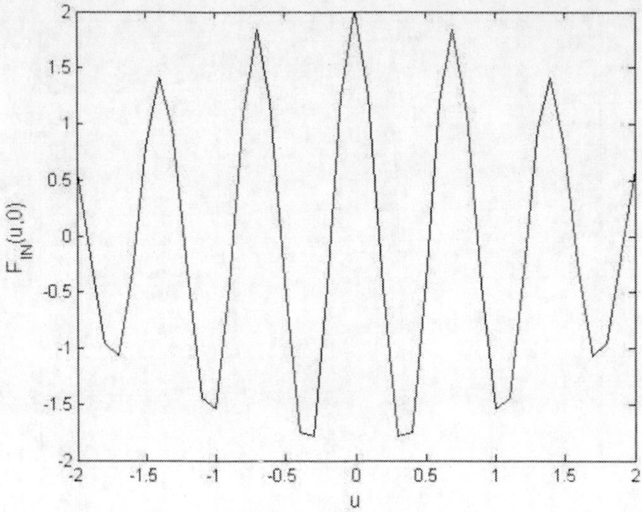

(a)

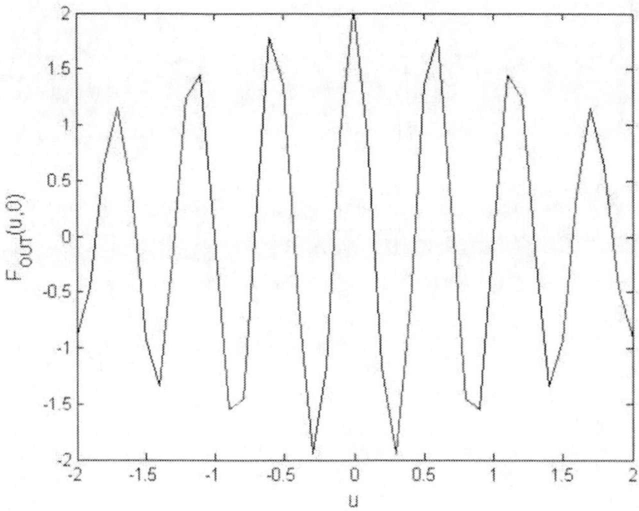

(b)

Figure 15 Axial (u, 0) distribution of the spectral pattern for (a) inward and (b) outward arrowhead without shaft.

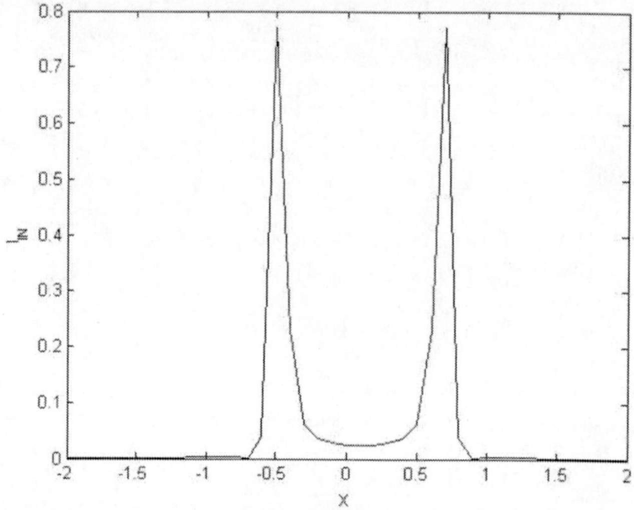

(a)

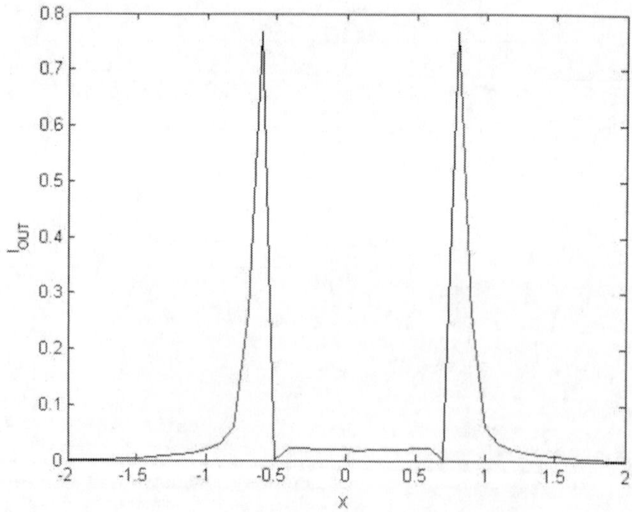

(b)

Figure 16 Image intensity distribution (a) $I_{IN}(x,0)$ and (b) $I_{OUT}(x,0)$ for inward arrow and outward arrow without shaft respectively.

Conclusion

The image analysis and processing system of vision is much involved. So, it may not be very simple to predict a particular model of eye which explains all the visual processing. It is seen that though most of the observations in the visual process point towards the fact that an image is formed on the retina but the recognition process, can not be accommodated in such a simple model. Many models were suggested by different vision scientists and these methods were tested with the explanation of different kinds of illusions. There are many explanations regarding the different illusion problems but these explanations do not lead us to finite goal. Haffenden and Goodale gave an explanation of the Ebbinghaus illusion that a central circle appears smaller when surrounded by large circles than when surrounded by small circles, using dorsal and the ventral stream theory [27]. However, Franz et. al. [28] after some experiments contradicted this claim. The purpose of this article is to suggest some model of human visual system in which the processing is taking place in the spectrum plane. Such processing was suggested by different authors as mentioned in the earlier section. Weale [29] explained the Orbison's, Luckiesh's and Zollner's illusions, which are versions probably of a single phenomenon, using the concept that lens is a low pass spatial filter. In this article we have explained Muller-Lyer illusion with the same concept of low pass filter in details so that it helps the interested scientific workers do venture into the field. It is felt that a systematic study on imaging mechanism with different types of spatial filters that is, selecting different zones of spectrum distribution might be applicable to explain other kinds of illusion, which in turn will enrich us about the response properties of retinal ganglion cells. Moreover, retinal receptors are not linear in nature regarding sensing of light intensity and this may cause some further complexity in visual processing system. All these aspects need a careful study for establishing a coherent logic in visual cognition process.

References

Albert, M. K. (1993). "Parallelism and the perception of illusory contours" *Perception*, *22*, 589-595.

Barlow, H. R. & Levick, W. R. (1969). "Three factors determining the reliable detection of light by retinal ganglion cells of cat", *J. physiol. London*, *200*, 1-24.

Basuray, A. & shah, M. (1991). "OTF evaluation utilizing a rotationally symmetric sinusoidal grating and a sinusoidal grating of finite height" *Jour' Opt. Soc. Am.(A) 8*, 212-217.

Campbell, F. W. & Robson, J. G. (1968). "An application of Fourier analysis to the visibility of gratings", *Jour. Of Physiology*, *159*, 551-568.

Duffieux, P. M. (1946). "L'integrale de Fourier et Ses Applications a L'optique" Fac. Sci, *Besancon, France*.

Enroth-Cugel, C. H. & Robson, J. G. (1966). "The contrast sensitivity of retinal ganglion cells of cat" *J. Physiol. London*, *187*, 517-552.

Franz, V. H., Gegenfurtner, K. R., Bulthoff, H. H. & Fahle, M. (2000). "Grasping visual illusion: No evidence for a dissociation *between perception and action*" *11*, 20-25.

Glennerster, A. & Rogers, B. J. (1993). New depth to the Muller-Lyer illusion. *Perception*, *22*, 691-704.

Glezer, V. D., Yakovlev, V. V. & Gauzelman, V. E. (1989). " Harmonic basic functions for spatial coding in the cat striate cortex". *Visual Neuroscience*, *3*, 351-363.

Goodale, M. A. & Milner, A. D. (1992). "Separate visual pathways for perception and action." *Trends in Neurosciences*, *15*, 97-112.

Graham, N. (1988). *Visual pattern analyzers*", New York: Oxford University Press.

Haffenden, A. M. (1998). "The effect of pictorial illusion on prehension and perception" *Journal of cognitive Neuroscience*, *10*, 122-136.

Hampson, K. M. (2008). "Adaptive optics and vision" *Journal of Modern Optics*, *55*, 3425-3467.

Hopkins, H. H. (1952). "On the diffraction theory of optical images," *Proc. Royal Soc. Series*, A, *217*, 408-432.

Kabrisky, M. (1966). *A proposed model for information processing in the human brain*", Urbana: University of Illinois press.

Kaisza, G. (1955). "Margini Quasi-persettivi in campi con stimulaione omogenea." *Rivista di pcicologia*, *49*, 7-30.

Kravstov, B. "A spectral model of visual perception" http://kravsoft.com/Spectral_Model.htm.

Land, M. F. & Nilsson, D. E. (2009). Animal eyes, Oxford University press.

Maffei, L. & Fiorentini, A. (1973). "The visual cortex as a spatial frequency analyzer" *Vision Research*, *13*, 1255-1267.

Parker, Andrew & Hawken, (1985). "Mike Capabilities of monkey cortical cells in spatial resolution tasks" *Jour. Opt. Soc. Am. A*, *2*, 1101-1114,.

Pollen, D. A. & Lee, G. R. (1971). "How does the strite cortex begin reconstruction of the visual world" *Science*, *173*, 3991, 74-77.

Ray. Mina (Shah), (1999). Basuray, Amitabha and Ray, Baidyanath. "Functional implications of spatial filtering on optical illusion" Optik, 110, 113-117.

Sternberg, (2003). Robert, *Cognitive Psychology* Third Edition, Thomson Wadsworth,.

Ungerleider, L. G. & Mishkin, M. "Two cortial visual system", In: D. J. Ingle, M. A. Goodale, & R. J. W. Mansfield (Eds.), *Analysis of visual behavior*, Cambridge: MIT Press.

Uttal, R. L., Kakarala, R., Dayanand, S., Shepherd, T., Kalki, J., Lunskis, C. F. & JR. Liu, Ning, (1999). *Computational modeling of vision*", Marcel Dekkar,.

Von Helmholtz, H. (1962). *Popular Scientific Lectures*; Dover: New York,.

Weale, R. A. (1977). "Optical illusions and visual functions"., In: S. J. Cool, & Smith, E. L. (Eds.), *Frontiers of visual science* Springer-Verlag, New york, 308-315.

Westheimer, G. (1960). "Modulation thresholds for sinusoidal light distribution on the retina." *J. Physiol.*, *152*, 67-74.

Wilson, H. R. & Gelb, D. J. (1984). "Modified line element theory for spatial frequency and width discrimination", *Jour. Opt. Soc. Am.*, *1*, 124-131.

In: Spatial Memory
Editor: Jacob B. Thomas, pp. 169-177 ISBN: 978-1-61668-139-5

Chapter 7

EGOCENTRIC CODING OF TARGET LOCATION IN VISUO-SPATIAL WORKING MEMORY

Sergio Chieffi[1*], Alessandro Iavarone[2-3], Giovanni Messina[1], Marcellino Monda[1] and Sergio Carlomagno[4]

[1]Department of Experimental Medicine, Section of Physiology, Second University of Naples, Italy
[2]Neurological and Stroke Unit, CTO Hospital, Naples, Italy
[3]Department of Relational Sciences, University of Naples "Federico II"
[4]Department of Psychology, University of Trieste, Italy

SUMMARY

Primate and human spatial location coding is based on a variety of different reference frames. Two fundamentally different types of reference system are usually considered, egocentric and allocentric. The first type codes locations with reference to the position of the body, or of specific parts of the body, the second codes locations with reference to the surrounding

* Corresponding author: Dipartimento di Medicina Sperimentale, Sezione di Fisiologia, Via Costantinopoli 16, 80138 Napoli, Italy, E-mail address: sergio.chieffi@unina2.it, Fax number: +39 +81 +5667500

environment. Herein we examine behavioural and neurophysiological evidence that suggests: (a) the existence of a series of co-ordinate transformations, from a retinotopic to body-centred spatial representation, for guiding bodily actions in space, and (b) the use of a body-centred co-ordinate system for coding, in short-term memory, of target locations to be reached with arm movements.

Research on human memory suggests a modular, domain-specific organization, both in temporary memory and in processing dependent on long-term memory [1]. In the original formulation of their 'working memory' (WM) model, Baddeley and Hitch [7] postulated the existence of a unitary, limited-capacity 'visuo-spatial scratch-pad (VSSP) for temporary representation and manipulation of all visuo-spatial information. Other studies, using selective interference paradigms, suggested to fractionate the VSSP into two components, one 'visual', the other 'spatial' [24,25,41]. Within 'spatial' short-term memory, a further distinction, also based on selective interference effects, has been made between memory for the subject's own, egocentric body postures or movements, vs. exocentric positional coding of objects [34,35]. Finally, behavioural evidence suggests that memory for exocentric visual locations can encode the direction and distance of the target independently [11].

Spatial representation involves a multiplicity of different coding systems and our understanding of space is derived from the complex interactions between them. Two fundamentally different types of reference system are usually considered, egocentric [11,12,15,36,37,38,40] and allocentric [8,10,22,23,30]. The first type codes locations with reference to the position of the body, or of specific parts of the body, the second codes locations with reference to the surrounding environment, independently from the subject's location or viewpoint.

Egocentric representation of space is obtained by integrating, continuously, visual and proprioceptive information. Neurophysiological studies have shown a series of co-ordinate transformations, from a retinotopic to a head-centred and body-centred spatial representation. Early visual processing in the primate visual system appears to be coded in retinotopic co-ordinates [13]. However, a retinotopic representation of visual space is not appropriate for guiding bodily actions in space, since the position of the eyes and head changes continuously. More stable spatial representations, adapted to the needs of action planning and memory, are progressively elaborated in the posterior parietal cortex, prefrontal cortex and elsewhere. Posterior parietal cortex appears to be responsible for integrating many different egocentric frames of reference both within and between modalities, presumably to ensure that any one location in space will be coded as the same location across all modalities [3,4,6]. A head-centred co-ordinate representation may be obtained by combining both the retinal location of a visual stimulus and information signalling the position of the eye [5], while a body-centred representation may be reached by combining the retinal and eye position signals with head position information [9]. Coding systems apparently centred on the face, trunk, arm, and other body parts have been reported in inferior premotor cortex, or area F4 [16,17,20,21]. Area F4 receives a major input from the inferior parietal lobule [26,31]. Graziano et al. [20,21] found neurons in ventral premotor cortex and in Putamen that responded to visual stimuli in the space adjacent to the hand or arm. Furthermore, the visual receptive fields of these neurons moved when the arm moved. The authors [20,21] suggested that these cells code space in arm-centred co-ordinates.

To isolate a particular spatial coding system experimentally, it is essential to constrain the subject's task in such a mode as to reduce (if possible, to eliminate) the use of all alternative coding strategies. The study of memory for a visually presented, egocentric target location requires that the information about spatial relations between target location and other, extrinsic reference objects is rigorously excluded.

Soechting and Flanders [36,37] asked the subjects to point to a remembered location of a target presented visually. They found that pointing errors were more consistently centred on the shoulder than on the eyes or head. However, in a subsequent study, Soechting et al. [38] showed that subjects may use a head-centred reference frame, when asked to point to locations relative to the head, and a shoulder-centred reference frame, when asked to point to locations relative to the shoulder. Flanders et al. [15] proposed a model in which the spatial representation of a target location for manual pointing evolves from a retinocentric to a head-centred and shoulder-centred representation.

Behavioural evidence for a hand-centred representation was found by Tipper et al. [40]. The authors [40] asked their subjects to reach for a target stimulus while avoiding an irrelevant distractor. When the distractor was presented in the space near the hand, it produced interference effects (and 'negative priming') dependent on its relation to the starting position of the hand, irrespective of its visual (i.e., view-point-dependent) position. Evidence for a hand-centred co-ordinate system in the planning of simple, 2-D reaching movements has been also reported by Ghez et al. [18] and Gordon et al. [19].

Chieffi et al. [12] found that visual target locations may be encoded in a hand-centred frame of reference in spatial short-term memory. The authors [12] required subjects to remember (over delays of 3 and 30 s) the location of a single, visually presented stimulus and to recall its location by pointing to it with eyes closed. Stimulus presentation was made either in the dark, providing purely egocentric visual information, or in the light, in which both egocentric and allocentric information was available. Pointing to the remembered location, however, was done with eyes closed in both cases, that is without allocentric visual guidance. There were two starting positions for the hand: one near the body, proximal to the target area, the other far from the body, i.e. beyond the target area. The results showed a large and consistent undershoot of the remembered target location with respect to the hand's starting position both when the hand started from the near and from the far position. Further, undershoot error in the remembered target location increased over time delay. Both these results were consistent with the hypothesis that visual target locations may be encoded in a hand-centred frame of reference in spatial short-term memory and suggested, further, the use of motor- or action- related representations in visuo-spatial working memory [12].

Direction and distance of egocentrically-specified visual location appear to be encoded independently in spatial short-term memory [11,23]. Chieffi and Allport [11] required subjects to remember the location of a briefly presented luminous spot. Stimuli were presented in the dark, thus providing purely egocentric spatial information. In Exp. 1, after the specified delay (4 or 30 s), subjects were instructed to point to the remembered location. The results showed that memory for both the direction and the distance of the visual target location were essentially uncorrelated, at all delay, and declined over time. In Exp. 2 the subjects were required to remember the locations of two visual stimuli presented sequentially, one after the other. The results showed that the encoding of a second location selectively interfered with memory for the direction but not for the distance of the to-be-remembered target location. The authors [11] suggested that the data of both the experiments provided evidence for a dissociation between memory for the distance and direction of a spatial location to be pointed to. The independent coding of the two parameters in spatial working memory parallels that found for immediate planning of movement distance and direction [14,19,32] and is strongly suggestive of the use of a motor-related representation in spatial working-memory. However, this does not imply that all spatial representation in short-term memory is coded in terms of motor plans, nor that the latter are related only to hand movements. Representation in terms of other sorts of planned movements (locomotor, oculo-motor, whole body postures, lip and tongue movements, etc.) may all contribute to our spatial representational abilities [11]. Recently, Lemay and Stelmach [23] found further evidence for the independent coding of direction and distance of the to-be-remembered target location. The authors [23] asked subjects to locate, memorize, and point to a target in a dark environment. However, during the recall delay the subject had to move their hand or their eyes. Lemay and Stelmach [23] observed an increase in variability in the directional component of the movement when the participants were required to move their hand during recall delay, and an increase in variability in the amplitude component when they were required to move their eyes. The authors [23] concluded that the direction of the movement is coded and remembered in a frame of reference linked to the arm, whereas the amplitude of the movement is remembered in an eye-centered frame of reference.

Position of exocentric object locations can be determined using, other than an egocentric frame of reference, also an allocentric frame of reference. In this case, target position is determined in relation to stable visual landmarks surrounding the target, independently of any egocentric reference [8,30]. Neurons in the hippocampus ('place' cells) have been found to fire when an animal passes through a given location in the environment regardless of the direction in which the animal crosses the location [28]. Place cells have been also observed in the hippocampus of primates [29]. Allocentric representations may depend on a direct correspondence between absolute positions in space and the activity of certain regions or neurons in the brain, allowing allocentric neural map of absolute space to be derived without reference to the mechanisms underlying egocentric coding. Alternatively, it is possible the existence of a continuum of codes from purely retinotopic, through more egocentric to allocentric representation, rather than a clear dichotomy between egocentric and allocentric coding. The integrative activity of the posterior parietal cortex might provide a sufficient basis for emergent allocentric representation [39]. However, the possibility remains that egocentric and allocentric coding systems in humans have become specialized to such a degree that they are at least functionally dissociable [33].

Woodin and Allport [42] demonstrated a clear double dissociation between memory for a target location fixed with respect to the stationary environment ('allocentric' location) vs. memory for a target location fixed with respect to the subject's body ('egocentric' location), where the environmental reference frame was provided only by vestibular and proprioceptive cues. In their experiment subjects, seated in a rotating chair with closed eyes, were required to point to remembered, auditorily presented target location. The subjects were rotated by the experimenter between the presentation and recall of targets. The target either was held stationary as the chair moved, preserving its allocentric position but not its egocentric position, or moved in tandem with the chair, preserving its egocentric position. The rotation of the chair was either followed by one of two types of interference task or by an unfilled delay. In both types of interference task the subject had to point, on verbal prompt, to a series of pre-learnt locations. In one task, the allocentric interference task, the pre-learnt locations remained fixed as the chair rotated whereas in the other task, the egocentric interference task, they rotated with the chair. Finally, the subject had to point to the target location. The results showed that memory for the target locations was selectively impaired when distractor locations were specified within the same spatial reference frame as the target. In contrast, distractor locations specified in a different reference frame from that of the target had either little or no effect on memory for target locations.

When both allocentric and egocentric coding are possible, tasks demands seem important in determining whether an allocentric [22] or both egocentric and allocentric strategies are used [10]. Lemay et al. [22] compared the kinematics and accuracy of a pointing movement to a remembered target when egocentric, allocentric, or combined egocentric and allocentric coding was possible. Subjects were required to localize, memorize, and reach to a target presented (a) in a completely dark environment (i.e., egocentric coding), or within a visual context provided by an illuminated square that (b) remained at the same position (i.e. egocentric and/or allocentric coding) or (c) was shifted to a different location (i.e., allocentric coding). Results indicated that pointing to a remembered allocentric target was more stable than that to a remembered egocentric target. Further, when both allocentric and egocentric coding were possible, the allocentric strategy was preferentially used. In their study, Carrozzo et al. [10] instructed participants to point to a remembered target position and informed them that although target location varied from trial to trial it could fell along a "virtual" line. The results showed that the pattern of variable errors could be accounted for egocentric and allocentric reference frames. The authors [10] concluded that target information can be defined in both egocentric and allocentric frames of reference, resulting in two independent and coexisting representations.

REFERENCES

Allport, DA. Auditory-verbal short-term memory and conduction aphasia. In: Bouma H, Bouwhuis DG editors. *Attention and Performance*. X. Hillsdale: Erlbaum; 1984, 313-325.

Andersen, RA. Inferior parietal lobule function in spatial perception and visuomotor integration. In: F; Plum, VB; Mountcastle, editors. *Handbook of physiology. The nervous system. Higher functions of the brain*. Rockville, MD: American Physiology Society; 1987, 483-518.

Andersen, RA. Coordinate transformations and motor planning in posterior parietal cortex. In: MS; Gazzaniga, (Eds.), *The Cognitive Neurosciences*. Cambridge, MA: MIT Press; 1995, 519-532.

Andersen, RA. Multimodal integration for the representation of space in the posterior parietal cortex. *Philos Trans R Soc Lond B Biol Sci*, 1997, 352,1421-8.

Andersen, RA; Essick, GK; Siegel, RM. Encoding of spatial location by posterior parietal neurons. *Science*, 1985, 230, 456-8.

Andersen, RA; Snyder, LH; Bradley, DC; Xing, J. Multimodal representation of space in the posterior parietal cortex and its use in planning movements. *Annu Rev Neurosci*, 1997, 20, 303-30.

Baddeley, A; Hitch, G. Working memory. In: G; Bower, (Eds.), *The Psychology of Learning and Motivation,* Vol VII. New York: Academic Press; 1974, 47-90.

Bridgeman, B. Separate visual representations for perception and for visually guided behavior. In: Ellis SR, editor. *Pictorial communication in virtual and real environments*. London: Taylor & Francis; 1991, 316-327.

Brotchie, PR; Andersen, RA; Snyder, LH; Goodman, SJ. Head position signals used by parietal neurons to encode locations of visual stimuli. *Nature*, 1995, 375, 232-5.

Carrozzo, M; McIntyre, J; Zago, M; Lacquaniti, F. Viewer-centered and body-centered frames of reference in direct visuomotor transformations. *Exp Brain Res*, 1999, 129, 201-10.

Chieffi, S; Allport, DA. Independent coding of target distance and direction in visuo-spatial working memory. *Psychol Res*, 1997, 60, 244-50.

Chieffi, S; Allport, DA; Woodin M. Hand-centred coding of target location in visuo-spatial working memory. *Neuropsychologia*, 1999, 37, 495-502.

DeYoe, EA; Van Essen, DC. Concurrent processing streams in monkey visual cortex. *Trends Neurosci*, 1988, 11, 219-26.

Favilla, M; Hening, W; Ghez, C. Trajectory control in targeted force impulses.

VI. Independent specification of response amplitude and direction. *Exp Brain Res*, 1989, 75, 280-94.

Flanders, M; Helms Tillery, SI; Soechting, JF. Early stages in a sensorimotor transformation. *Behav. Brain Sci*, 1992, 15, 309-62.

Fogassi, L; Gallese, V; di Pellegrino, G; Fadiga, L; Gentilucci, M; Luppino, G; Matelli, M; Pedotti, A; Rizzolatti, G. Space coding by premotor cortex. *Exp Brain Res*, 1992, 89, 686-90.

Fogassi, L; Gallese, V; Fadiga, L; Luppino, G; Matelli, M; Rizzolatti, G. Coding of peripersonal space in inferior premotor cortex (area F4). *J Neurophysiol*, 1996, 76, 141-57.

Ghez, C; Gordon, J; Ghilardi, MF; Sainburg, R. Contributions of vision and proprioception to accuracy in limb movements. In: Gazzaniga MS, editor. *The Cognitive Neurosciences*. Cambridge, MA: MIT Press; 1995, 549-64.

Gordon, J; Ghilardi, MF; Ghez, C. Accuracy of planar reaching movements. I. Independence of direction and extent variability. *Exp Brain Res*, 1994, 99, 97-111.

Graziano, MSA; Gross, CG. Multiple pathways for processing visual space. In: T; Inui, JL; McClelland, editors. *Attention and Performance XVI*. Cambridge, MA: MIT Press; 1996, 181-207.

Graziano, MS; Yap, GS; Gross, CG. Coding of visual space by premotor neurons. *Science*, 1994, 266, 1054-7.

Lemay, M; Bertram, CP; Stelmach, GE. Pointing to an allocentric and egocentric remembered target. *Motor Control*, 2004, 8, 16-32.

Lemay, M; Stelmach, GE. Multiple frames of reference for pointing to a remembered target. *Exp Brain Res*, 2005, 164, 301-10.

Logie, RH. *Visuo-Spatial Working Memory*. Hove, UK: Lawrence Erlbaum Associates; 1995.

Logie, RH; Marchetti, C. Visuo-spatial working memory: visual spatial or central executive? In: RH; Logie, M; Denis, (Eds.), *Mental images in human cognition*. Amsterdam: North Holland Press; 1991, 105-15.

Matelli, M; Camarda, R; Glickstein, M; Rizzolatti, G. Afferent and efferent projections of the inferior area 6 in the macaque monkey. *J Comp Neurol*, 1986, 251, 281-98.

McIntyre, J; Stratta, F; Lacquaniti, F. Short-term memory for reaching to visual targets: psychophysical evidence for body-centered reference frames. *J Neurosci*, 1998, 18, 8423-35.

O'Keefe, J; Nadel, L. *The hippocampus as a cognitive map*. Oxford: Clarendon; 1978.

Ono, T; Nakamura, K; Fukuda, M; Tamura, R. Place recognition responses of neurons in monkey hippocampus. *Neurosci Lett*, 1991, 121, 194-8.

Paillard, J. Motor and representational framing of space. In: J; Paillard, (Eds.), *Brain and space*. Oxford: Oxford University Press; 1991, 163-82.

Petrides, M; Pandya, DN. Projections to the frontal cortex from the posterior parietal region in the rhesus monkey. *J Comp Neurol*, 1984, 228, 105-16.

Rosenbaum, DA. Human movement initiation: specification of arm, direction, and extent. *J Exp Psychol Gen*, 1980, 109, 444-74.

Sherry, DF; Schacter, DL. The evolution of multiple memory systems, *Psychol Rev,* 1987, 94, 439-54.

Smyth, MM; Pearson, NA; Pendleton, LR. Movement and working memory: patterns and positions in space. *Q J Exp Psychol A*, 1988, 40, 497-514.

Smyth, MM; Pendleton, LR. Space and movement in working memory. *Q J Exp Psychol A*, 1990, 42, 291-304.

Soechting, JF; Flanders, M. Sensorimotor representations for pointing to targets in three-dimensional space. *J Neurophysiol*, 1989, 62, 582-94.

Soechting, JF; Flanders, M. Errors in pointing are due to approximations in sensorimotor transformations. *J Neurophysiol*, 1989, 62, 595-608.

Soechting, JF; Tillery, SIH; Flanders, M. Transformation from head- to shoulder-centered representation of target direction in arm movements. *J Cogn Neurosci*, 1990, 2, 32-43.

Stein, JF. Space and the parietal association areas In: Paillard J, editor. *Brain and space*. Oxford: Oxford University Press; 1991, 185-222.

Tipper, SP; Lortie, C; Baylis, GC. Selective reaching: evidence for action-centered attention. *J Exp Psychol Hum Percept Perform*, 1992, 18, 891-905.

Tresch, MC; Sinnamon, HM; Seamon, JG. Double dissociation of spatial and object visual memory: evidence from selective interference in intact human subjects. *Neuropsychologia*, 1993, 31, 211-9.

Woodin, ME; Allport, A. Independent reference frames in human spatial memory: body-centered and environment-centered coding in near and far space. *Mem Cognit*, 1998, 26, 1109-16.

In: Spatial Memory
Editor: Jacob B. Thomas, pp. 179-202
ISBN: 978-1-61668-139-5

Chapter 8

THE EFFECT OF VISUO-SPATIAL ATTENTION ON LONG-TERM MEMORY ENCODING

Stephan Josef Stegt[1*] and Cristina Massen[2]
[1]University of Bonn, Bonn, Germany
[2]Max-Planck-Institute for Human Cognitive and Brain Sciences, Leipzig, Germany

ABSTRACT

Existing studies about the effects of attention on long-term memory encoding focused on the effects of 'frontal', executive attentional mechanisms and made use of paradigms of divided attention. We investigated the impact of visuo-spatial ('posterior') attention on long-term memory encoding. In four experiments, we used a standard spatial cuing paradigm and presented words either under valid or invalid cuing conditions. Performance in yes-no-recognition, cued recall and free recall was enhanced in the valid cuing-condition, whereas no effect was found for source-monitoring. The results show not only divided, but visuo-spatial attention considerably influences memory encoding, and the impact of visuo-spatial attention is different from that of divided attention.

[*] Corresponding author: Institut für Psychologie der Universität Bonn, Römerstr. 164, D-53117 Bonn. Email: Stephan.Stegt@web.de

INTRODUCTION

Research about the role of attentional processes for encoding into long-term memory is important for the development of our theoretical understanding of memory, but has neglected a differentiation of the concept of attention, focusing mainly on divided attention (DA) (e.g. Baddeley, Lewis, Eldridge & Thomson, 1984; Craik, Govoni, Naveh-Benjamin & Anderson, 1996). For this reason, we investigate the impact of visuo-spatial attention (VSA) on encoding processes.

It is a widely accepted assumption that memory consists of several distinctive components, called systems or processes that work together. The crucial issue in contemporary memory research is to identify these different components and reach agreement about which processes take place when we encode or retrieve information. In this research agenda, one of the key issues is to find out whether different memory components might be differentially affected by manipulations of attention. In fact, looking for effects of attention on performance in different memory tests has often been a tool to detect and describe memory systems (Jacoby, Woloshyn & Kelley, 1989; Gardiner & Parkin, 1990; Mulligan, 1998; Fernandes & Moscovitch, 2000).

During the last decade of the 20th century, the view of attention as a unique and limited resource has been abandoned in favour of different domain-specific mechanisms of attention (e.g. Allport, 1993; Neumann, 1992). Allport (1993) differentiates non-spatial and spatial attention mechanisms. Following Allport, non-spatial attentional processes can be seen as aspects of executive control. These processes can be investigated in dual-task studies. Many researchers did so and investigated these processes and their impact on memory encoding. Spatial attentional processes are important for the selection of stimuli to control one's actions by amplifying the processing of stimuli or particular features of stimuli. They are typically investigated in paradigms of selective attention, for example the spatial cuing paradigm, where a cue directs attention to a spatial position before a target item appears. It has been shown that subjects react more quickly to target items presented at cued positions within the visual field (Posner, 1980). Visual attention accelerates early perceptual processes and a consciousness threshold is reached more quickly (Alexander & Reinitz, 2000).

Nevertheless, other findings suggest that visual attention affects not only early perceptual, but also higher-level processes. This has been shown especially for a paradigm termed inhibition of return (IOR), in which attention is first directed away from the fixation point by a peripherally presented cue, and then redirected towards the fixation point. After these events, the return of attention to the previously cued location seems to be inhibited. Some studies show interactions between IOR and higher cognitive processes. Chasteen and Pratt (1999) found, that IOR does not only inhibit returning attention to a previously attended location, but also affects the higher-level processing that is required for lexical access, which was revealed by an interaction of IOR and word frequency on lexical decision times. Fuentes, Vivas and Humphreys (1999) presented prime words, which were later followed by a target word, either under conditions of IOR or not. For the target word, a lexical decision had to be made by the subject. The target was either semantically related or unrelated to the prime word. When the prime word had been presented under conditions of IOR, reaction times for semantically related target words were considerably slowed. Fuentes et al. (1999) proposed as an explanation that this negative priming stems from a temporary inhibitory tagging of the stored cortical representations of the primes in semantic memory, when the primes are presented under conditions of IOR. This inhibition could spread to semantically related items.

The results cited above suggest, that VSA does not only affect early perceptual processes, but also processing of items at later stages. Moreover, there are neuroimaging findings indicating that the posterior neocortex, which is implicated in VSA, could also be involved in memory encoding (Mayes & Montaldi, 1999; Otten, Henson & Rugg, 2002; Henson, Hornberger & Rugg, 2005).

Summing up the presented results, it emerges as an important question if memory processes are also affected by posterior VSA, and how they are affected. As the former research on attention and memory was mainly restricted to non-spatial, frontal attention and dual task studies, it is our concern to investigate the role of spatial, non-executive attention for memory encoding. Some studies already suggest that the impact of attention on memory *retrieval* does not only depend on the kind of memory test used, but also on how attention is manipulated. For DA, it has been shown that its manipulation during retrieval does not influence performance in yes-no recognition tests (Baddeley et al., 1984; Craik et al., 1996), but this statement does not hold for VSA. Massen and Stegt (2007) showed that the manipulation of inhibition of return at test impairs recognition decisions. They conducted four experiments in which word or nonword stimuli were learned under different encoding conditions (e.g. with a levels-of-processing manipulation). At test, old items and distractors were presented either under conditions of inhibition of return or not. Performance was worse under inhibition of return (fewer hits and more false alarms), but only if the recognition decision was difficult, as is the case for nonwords, for high-frequency words, or for shallowly encoded words. Interestingly, this it what some authors (Naveh-Benjamin, Craik, Perretta & Tonev, 2000) expected for a DA manipulation at test: a reduced performance should only be found for difficult retrieval conditions. But these authors found no support for this assumption. The results of Massen and Stegt show, that VSA affects memory access during retrieval. It is revealed again, that higher-level processes are affected by manipulations of VSA. In addition, the results show, that there might be important differences between the impact of frontal, executive attention on memory and the impact of VSA.

To our notion, there is only one study that uses direct memory tests and investigates the impact of selective attention on memory encoding. In the study of Bentin, Moscovitch and Nirhod (1998), the authors manipulated selective attention in their experiments 2 and 3 and measured performance in an explicit test (recognition) and two implicit tests (lexical decision and living/nonliving-decision). In their selective attention manipulation, two words were presented together at different spatial positions and in two different colors. Depending on the color of a rectangle placed between the two stimuli, the subject had to process one word and to ignore the other. The recognition test revealed a marked advantage for the attended words. However, although these authors used a selective attention paradigm, it seems likely that executive attentional processes influence memory encoding in this object-centered selective attention paradigm. This is, because a distractor word was present during encoding, which had to be suppressed, while the target word was actively selected. According to Allport (1993), the active inhibition or suppression of task-irrelevant stimuli or stimulus features is a nonspatial, executive control process.

Thus, we decided to use the classical spatial cuing paradigm (Posner & Cohen, 1984), which might be a more pure realisation of the spatial attentional system proposed by Allport (1993), to study the impact of selective attention on memory encoding. In such a paradigm, no distractor is present during encoding and with explicit encoding instructions, the amount of frontal resources attributed to each item will be equal, such that any observed effects are due to the manipulation of spatial attention, rather than to differences in attributed frontal processing resources.

Divided Attention and Memory

Research about divided attention (DA) and memory has been central for two major theoretical developments, namely for the dual component-approaches to memory with the distinctions of explicit / implicit memory or automatic / controlled processes, and also for the neuropsychological memory model of Moscovitch (1992).

Dual component-approaches to memory posit a qualitative distinction between two kinds of memory or memory processes, often termed recollection vs. familiarity, controlled vs. automatic processes or explicit vs. implicit memory. These distinctions are closely related to the assumption that attention is important for controlled or recollective processes whereas automatic or familiarity processes do not rely on attention. A series of studies that used the dual-task paradigm supported and developed that idea (e.g. Gardiner & Parkin, 1990; Mulligan, 1998; Jacoby, Toth & Yonelinas, 1993). Gardiner and Parkin (1990) conducted an experiment in which they compared undivided attention at study with two DA conditions, which differed from each other with respect to the difficulty of the concurrent task. The test was a recognition test that made use of the *remember-know* distinction suggested by the work of Tulving (1985). Dividing attention during encoding progressively impaired the amount of conscious recollection ('remember') of the study episode, but recognition in the absence of recollection ('know') was not affected. The authors concluded that their results provide support for the assumption that recognition consists of two distinct components, one based on associative and contextual information, the other based on an awareness of familiarity.

Several authors proposed that dividing attention during encoding dissociates performance on implicit and explicit tests. Jacoby, Woloshyn and Kelley (1989) reported that dividing attention during encoding reduced explicit test performance but left repetition priming unaffected. Similar findings were obtained by other researchers (Russo & Parkin, 1993). Mulligan (1998) had a closer look on this assumption and reports a series of 5 experiments in which attention was divided during the study phase. He used four kinds of memory tests: a perceptual implicit test (word-fragment completion), a perceptual explicit test (word fragment cued recall), a conceptual implicit test (word association), and a conceptual explicit test (associate-cued recall). Dividing attention impaired performance in all but the perceptual implicit tests. The important point of his research was that performance on a conceptual implicit test, the word association task, was influenced by DA.

Finally, Jacoby, Toth and Yonelinas (1993) performed dual-task studies to underline the validity of their process-dissociation procedure designed to measure the contribution of conscious recollection versus unconscious, automatic processes. In their experiments 1a and 1b, they showed that dividing attention did not influence automatic processes, but had an enormous effect on the recollection measure: it fully eliminated conscious recollection. To sum up, the proponents of dual-component approaches to memory conducted DA studies to support their hypotheses. While the problem of the exact nature of the two processes or components remains unresolved, the studies have been of major importance for our understanding of memory, and the procedures developed in this research have proven very fruitful in other domains of memory research.

The theoretical importance of attentional effects on memory performance is also revealed if one considers neuropsychological models of memory, like the one by Moscovitch (1992). In his modular model, he postulates four essential components of memory, which dominate performance on different types of memory tests. One component consists of several perceptual and semantic modules. These modules mediate performance on perceptual, item-specific tests, without conscious memory for the learning episode. The basal ganglia constitute a second component, mediating performance on sensomotoric, procedural memory tests. A third, hippocampal component is also modular and implicated in explicit, episodic memory. It is often called the MTL/H (medial temporal lobe/hippocampal) component and responsible for performance in direct tests of episodic memory as recognition, cued recall and free recall. A fourth component is called the strategic component and requires the prefrontal cortex (PFC). It is a central system under voluntary control, responsible for working with memory, the exchange between working memory and long-term memory, for source-memory decisions or strategic, explicit tests like free recall of categorized lists.

Following Moscovitch, the third component is domain-specific and operates in an automatic manner during encoding and retrieval. It binds or integrates the engrams of the modules and central interpretative systems, resulting in a memory trace for every (and only) consciously perceived information. This trace contains the content of the information as well as the features that represent the awareness of the learning episode. At test, a perceived cue can automatically trigger the hippocampal index and interact with the stored information, a process also called "ecphory" by Tulving (1983). Once initiated, such ecphoric processes are rapid and obligatory, and they lead to a conscious recollection of the event. Moscovitch postulates that in contrast to this third component, the fourth, frontal component works with restricted resources and is domain-independent, taking care that attention is directed to the to-remembered items and that the necessary organisation takes place. During retrieval, we need the frontal component only if search strategies or organisational processes are needed, because no effective retrieval cues are available.

The Moscovitch model makes predictions for the impact of attention on encoding and retrieval. On the one hand, a reduction of frontal resources by a secondary task during encoding should always, independently from the secondary task itself, lead to a reduction of performance in a direct, explicit memory test, because attention cannot be properly directed to the input for the third component. So any concurrent task that diverts frontal resources decreases the likelihood that information is encoded adequately by the MTL/H component. However, a reduction of performance by a concurrent task during retrieval is dependent on the type of the memory test and the question whether strategic components are needed in this test, as is the case for source judgements or organisational processes. If only associative and automatic combination of a cue and a learning episode is required, this can happen automatically and will not be affected by the secondary task.

Consistent with these predictions, Baddeley et al. (1984) found that performance is reduced by dividing attention during encoding, but the same attentional manipulation had almost no effect on retrieval. This suggests that retrieval is an automatic process and can proceed without frontal resources. Craik et al. (1996) further investigated the impact of DA on memory and supported the results of Baddeley et al. (1984). They conducted four experiments and manipulated DA during encoding and retrieval, making use of various memory tests with different retrieval difficulty. In free recall (experiment 1 and 2), cued recall (experiment 3) and recognition (experiment 4), they showed that DA at encoding always led to a marked decrease in performance, whereas the impact of DA at test was only marginal or even non-existent. The authors also showed that retrieval is not a process that needs no resources at all, as performance (reaction times) in the secondary task was reduced by DA. Rather, retrieval is obligatory in the sense that it is given priority over other tasks. Interestingly, the results of Craik et al. (1996) also suggest that this allocation of resources is not under conscious control of the subjects.

There is evidence that a reduction of performance by a secondary task during retrieval is only found in tests that require prefrontal resources, like in tasks where the subject has to reproduce categorized lists or to discriminate lists during recognition (Moscovitch, 1994). A study of Fernandes and Moscovitch (2000) added to this conclusion that a concurrent task could also interfere with the retrieval process when both compete for the same perceptual representational system. These results show in accordance with the neuropsychological model that the effect of DA on memory retrieval depends on the kind of memory test used, whereas it always has an effect on memory encoding, at least if performance in direct, explicit memory tests is considered. In addition, the more executive processing is needed during encoding, the larger the effect of DA seems to be. Troyer, Winocur, Craik and Moscovitch (1999) showed that performance in source memory for location was decreased more by a concurrent finger-tapping or visual reaction-time task, than item memory.

To summarize, the evidence for a differential impact of divided attention on different types of memory tests can be explained quite well by the Moscovitch model. However, it is less clear what kind of attention underlies the conscious encoding and retrieval of the third component. We hypothesize that spatial attentional processes might be important for the automatic, cue-triggered working of this third component.

THE PRESENT APPROACH

In the experiments reported in this chapter, we take a closer look at the impact of VSA on long-term memory encoding by using a standard spatial cuing paradigm. In the learning phase of the experiments, items appeared in one of two spatial locations on the screen. Shortly before the appearance of the item, a cue was presented, either in the same or in the other location. This cue summons VSA. Subjects were required to speak out aloud all items they were able to identify. After a short distractor interval, a memory test followed and we compared performance for items presented in the cued location with that for items presented in the uncued location, conditionalized upon correct perceptual identification of items. We used a yes-no recognition test (experiment 1), a cued recall test (experiment 2), a free recall test (experiment 3), and a source memory test (experiment 4).

EXPERIMENT 1

In experiment 1, words were presented in a standard spatial cuing paradigm with compatible and incompatible trials during encoding, and memory was later tested with a standard yes-no-recognition test.

Method

Participants. 40 students from the University of Bonn received course credit or were paid 3 euro for their participation.

Procedure. Subjects sat in a distance of 40 cm from the computer screen. Instructions to speak out aloud all words presented, and to learn them for a later memory test were given on the screen prior to the study phase. In the study phase, 56 unrelated nouns (4 buffer-items first, then 52 randomly chosen items out of a list of 104 unrelated nouns) were sequentially presented. 13 items were presented in each cuing (valid-invalid) x location (above vs. below the middle of the screen) condition. The sequence of trials was random. Figure 1 depicts the sequence of events in a trial of the study phase. Each trial began with an empty screen for 100 ms, followed by a fixation cross in the middle of the screen, which was presented for 900 ms. Then the fixation cross disappeared and a box (2.5 cm x 1.5 cm) was presented with its center 2.5 cm above or below the fixation cross. The box was deleted after 100 ms and the to-be-remembered word appeared for 150 ms, with its centre 2.5 cm above or below the fixation cross. Then the display with the fixation cross was presented again for 3000 ms. During this period, the subject had to articulate the item. Then the next trial began as described above. If cue and item appeared in the same location, the item was presented in the valid cuing condition, if cue and item appeared in opposite locations, the item was presented in the invalid cuing condition. All stimuli were presented in white on a dark grey background. Letters were approximately 4 mm large and 5 mm high, the font was arial.

The verbal output of the subject was recorded with a standard tape recorder. For each subject, the items and their position in the study list had been determined by a randomization before the experiment, and the experimenter had a list of these items. He rated if the verbal output corresponded to the study item. This allowed conditionalizing memory performance upon correct item identification.

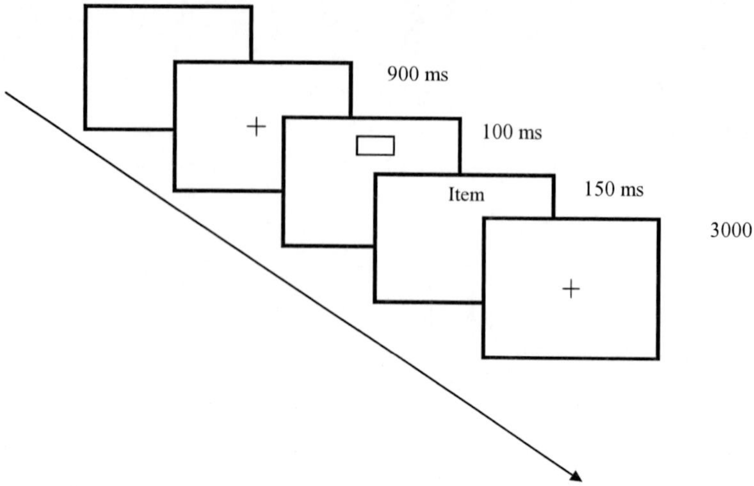

Figure 1. Sequence of events in a typical study trial of experiments 1 to 4. The figure shows the valid cuing condition. We presented words for only 150ms to discourage participants from making eye movements.

The study phase was followed by a 3-min distractor task, in which subjects had to solve arithmetic problems. In the test phase, the 52 items from the study phase, together with 52 distractor items, were presented sequentially in a random order in the middle of the computer screen. Subjects were required to press the 'j'-key (for the German word 'ja') when they recognized an item as old and the 'n'-key (for the German word 'nein') when they judged the item as new. Each test trial had to be initiated by the space bar. There was no time limit for the recognition decision.

Stimulus material and apparatus. The material consisted of 104 German four- and five-letter nouns. For each subject, 26 items were randomly assigned to the compatible cue-condition, 26 items to the incompatible-cue condition, and 52 were distractors for the recognition test. Presentation order was random in the learning and the test phase. The experiment was run on a pentium IV computer with a 17''-monitor and programmed with E-Prime (Schneider, Eschman & Zuccolotto, 2002).

Results

All statistical analyses in the present and the following experiments were done at the alpha = .05 level. For each test it is indicated individually if it was one-tailed or two-tailed.

Identification performance. An item of the study phase was considered as correctly identified, when the subject had been able to speak it out aloud. Plural forms of items presented in singular were accepted as correct identifications. Identification of presented items was better in the valid cuing condition, as had been expected. In the valid condition, the identification rate was .80, as compared to .64 in the invalid condition. This difference was significant with $t(39) = 4.9$, $SEM = .032$, $p < .001$, two-tailed.

Recognition performance. The false alarm rate was .20, the hit rates were .79 for the valid and .63 for the invalid condition. Performance was measured by the discrimination index ($Pr = H\text{-}FA$), following the two-high-threshold model of recognition memory, that has proven to be as valid as the signal detection analysis, but makes fewer theoretical assumptions (Snodgrass & Corwin, 1988). Performance was *Pr(valid)* = .59 vs. *Pr(invalid)* = .43. This difference was significant with $t(39) = 6.3$, $SEM = .024$, $p < .001$, two-tailed. Of course, there were less items in the invalid condition that were identified by the subjects, so recognition performance was conditionalized upon correct identification in the study phase for each participant. Recognition performance was still significantly higher in the valid condition with *Pr(valid)* = .68 and *Pr(invalid)* = .63, $t(39) = 2.8$, $SEM = .018$, $p < .01$, two-tailed. These results do not change when a signal detection analysis is done. Of the 40 subjects, 23 subjects showed superior performance for identified valid items, 6 showed no difference, and only 11 showed superior performance for identified invalid items.

Discussion

In Experiment 1, we tested the impact of a spatial cuing manipulation during encoding on performance in a recognition test. Recognition performance was better for items that were presented in the valid cuing condition, even when conditionalized upon correct identification. This result shows that not only frontal, executive attentional processes, but also processes of VSA affect long-term memory encoding.

One objection could be made to this conclusion. Following Moscovitch (1992), item recognition is a task that taxes the MTL/H component, but it has been shown that feelings of familiarity also contribute to old-responses in recognition, and that feelings of familiarity might be caused by perceptual fluency (Kelley & Jacoby, 1996; Lindsay & Kelley, 1996). When processing in perceptual modules is enhanced in the valid cuing condition, this could provoke an increase in familiarity of valid study items and lead to more 'old' responses for these items. Recall is also a test that taxes the MTL/H component of the Moscovitch model, but it is unlikely that familiarity due to perceptual fluency contributes to performance in this test. For that reason, a finding that the effect generalizes to recall tests would support our conclusion that VSA affects higher-level encoding processes beyond facilitating the processing in perceptual modules. Therefore, we conducted further experiments to test whether the results generalize to cued and free recall.

It might also be objected that the advantage of cued items results from a greater allocation of processing resources to these items. When less time is needed for item identification in the valid cuing condition, more time might be available for encoding processes. This hypothesis seems unplausible however, because the processing time advantage for items in the valid cuing condition is presumably very low, and because there are 3000 ms to pronounce and rehearse each item after stimulus presentation. In addition, the argument of an advantage for valid items due to processing resources is strongly challenged by studies on perceptual interference (Nairne, 1988; Mulligan, 1996). In these studies, items in the interference condition are presented shortly and then masked. Although they are more difficult to identify, they are better encoded into memory. In perceptual interference studies, the experimental conditions are similar to our experiments with respect to presentation duration and identification difficulty, but the items that are more difficult to identify are better remembered, contrary to the results in our experiment.

Nevertheless, to rule out any alternative explanation based on processing time, we increased rehearsal time for items in the invalid condition in experiment 2 and conducted a cued recall test.

EXPERIMENT 2

In experiment 1, a yes-no-recognition test revealed that items presented in the valid cuing condition at study are better encoded into memory than are items of the invalid cuing condition. Experiment 2 aimed to replicate these findings with a word stem cued recall test. A replication of the effect in cued recall would provide more convincing evidence for the notion, that VSA has an impact on long-term memory encoding mediated by the MTL/H component. In addition, providing more rehearsal time for items in the invalid condition rules out any alternative explanation based on processing time.

Method

Participants. A total of 40 students from the University of Bonn, who had not participated in experiment 1, participated. They received course credit or were paid 5 euro.

Procedure. The procedure in the learning phase and distractor task was identical to the procedure in experiment 1, with the exception that the duration of the last display of each trial in the invalid condition was increased from 3000 to 3200 ms.

At test, the first two letters of a test item were presented on the computer screen, and the subject had to complete the letters to a german five- or six-letter word. They were instructed to try to remember a corresponding word of the study phase and to complete to this word. If they did not remember any word of the study phase with these first two letters, they were to complete to the first german five- or six-letter word that came into their mind. They were given a maximum of 20 seconds to complete each word stem.

Stimulus material and apparatus. Word stimuli were 112 five- and six-letter nouns, taken from Krüger (1998). This change of stimulus materials was necessary for the cued recall procedure with word stem completion, as all of the Krüger items have different word stems. The other aspects were identical to experiment 1.

Results

Identification performance. As expected, significantly more items were identified in the valid cuing condition (.65) than in the invalid cuing condition (.46), $t(39) = 7.1$, $SEM = .026$, $p < .001$, one-tailed.

Cued recall performance. The word stem completions were rated by the first author and a psychology student. Orthographic errors were accepted, as well as plural forms of words presented in singular. For each individual item, we calculated its base rate for completion to a target word, when it had not been presented in the study phase. Then, we calculated the mean base rate for identified items of the valid cuing condition, and for identified items of the invalid cuing condition for each individual subject. There was no difference between the mean base rates of identified valid items and identified invalid items across participants (both .20). The recall rate was .37 for items presented in the valid condition and .30 for items presented in the invalid condition. This difference was significant, $t(39) = 4.0$, $SEM = .018$, $p < .001$, one-tailed. As there were again fewer items in the invalid condition that were correctly identified during study, completion rates were conditionalized upon correct identification. Completion rates for identified study items were .47 for the valid and .42 for the invalid condition. This difference reached statistical significance with $t(39) = 2.0$, $SEM = .025$, $p < .05$, one-tailed. When completion base rates for identified valid and invalid items for each subject are subtracted from the respective recall rates, this difference does not change, $t(39) = 1.9$, $SEM = .026$, $p < .05$, one-tailed. Of the 40 subjects, 27 (67.5 %) had a better performance for identified items of the valid cuing condition, and 13 had a better performance for identified items of the invalid cuing condition.

Discussion

Experiment 2 replicates the results of experiment 1 and demonstrates again that spatial attention affects memory encoding. Items that are presented under valid cuing conditions are better encoded into memory. If visual attention influenced only early perceptual processes, the first memory component in the Moscovitch model, the perceptual modules, would be affected. This could result in a higher familiarity for items in the valid cuing condition due to perceptual fluency, which in turn could lead to more old responses in a recognition test. But if this hypothesis was true, there would be no effect in recall. For that reason, the fact that the effect generalizes to cued recall strengthens the conclusion that VSA affects higher-level encoding processes, probably the MTL/H and/or the PFC component.

In addition, the alternative explanation of a processing time advantage for items in the valid condition is ruled out. Note that any advantage in rehearsal time for valid items would presumably be below 200 ms. To complete our understanding of the impact of VSA during encoding on episodic memory and to support the conclusions of experiments 1 and 2, experiment 3 investigates the impact of VSA in the third test that is meant to reflect the MTL/H component: free recall.

EXPERIMENT 3

Experiments 1 and 2 have shown that items that are presented under valid cuing conditions are better encoded into memory. In experiment 3, we used the same encoding procedure as in experiment 1 and tested whether VSA also affects performance in free recall.

Method

Participants. 32 students from the University of Bonn, who had not participated in any of the previous experiments, participated for course credit or 3 euro.

Procedure. The procedure was the same as in experiment 1 for the study phase and the distractor task. At test, subjects were given lined paper with the instruction to write down all items they had retained from the study phase. Subjects were given 5 minutes to recall.

Stimulus material and apparatus. Stimulus material and apparatus were identical to experiment 1.

Results

Identification performance. Again, significantly more items were identified in the valid cuing condition (.78) than in the invalid cuing condition (.64), $t(31) = 4.6$, $SEM = .031$, $p < .001$, one-tailed.

Recall performance. The recalled words were rated by the first author and a psychology student. Orthographic errors were accepted, as well as plural forms of words presented in singular. There were 1.1 intrusions per subject ($SD = 1.5$). More items of the valid cuing condition were recalled (recall rate = .17) than items of the invalid condition (recall rate: .12), $t(31) = 3.1$, $SEM = .018$, $p < .01$, one-tailed. Conditionalized upon identification performance, recall rates were still significantly higher for items that were presented in the valid cuing condition: .22 vs. .17, $t(31) = 1.8$, $SEM = .025$, $p < .05$, one-tailed. 21 subjects showed the predicted effect, and 11 subjects had an effect in the opposite direction. There were no ties. Figure 2 shows memory performance for valid and invalid cuing in experiments 1 to 3, conditionalized upon correct identification.

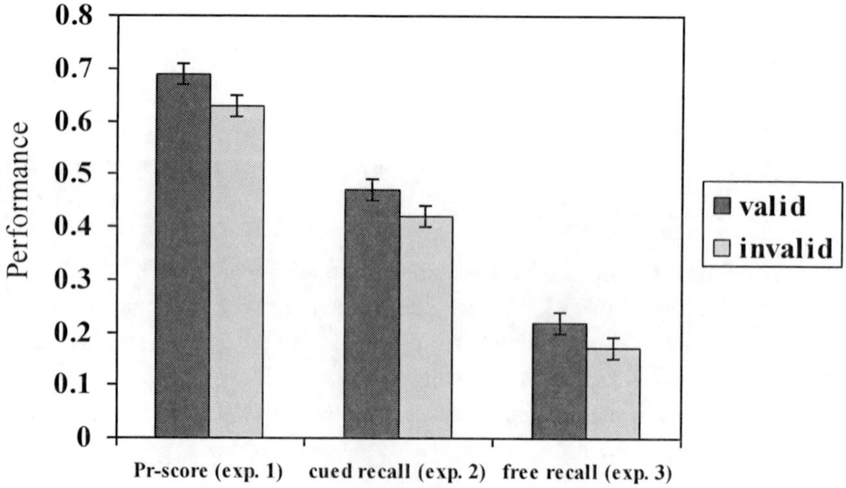

Figure 2. Recognition performance (Pr = H – FA) and recall performance (percent correctly recalled or completed) for the valid and invalid cuing condition in experiments 1 to 3. Measures are conditionalized upon correct identification. Error bars represent standard error of the mean difference.

Discussion

Experiment 3 replicates the findings of the first two experiments. It demonstrates again that spatial attention affects memory performance. Items that are presented under valid cuing conditions, are better encoded into memory. In three different memory tests, that are assumed to reflect the MTL/H component of the Moscovitch model, this impact of visual attention on performance can be found, whether performance is conditionalized upon identification or not. So far, the results resemble those obtained for manipulations of divided attention (e.g., Craik et al., 1996), although the effects seem to be smaller.

To find out to which extent this impact of VSA is parallel to that of DA, we now take a closer look at the fourth, the PFC component of the Moscovitch model. The PFC component mediates organizational, strategic processes during encoding and is thus especially important for source monitoring and/or recall of categorized lists. The effects of DA on memory are greater for source memory than for item memory (Troyer et al., 1999). It is interesting to work out, if the same is true for visuo-spatial attention. As both the PFC and the MTL/H component mediate performance in explicit tests, it is not sure whether the results of experiments 1 to 3 originate from an influence of VSA on the PFC, the MTL/H, or both components. If the PFC component is influenced by VSA, the effect should be stronger for source monitoring, as is the case for DA (Troyer et al., 1999).

EXPERIMENT 4

In this experiment, the impact of VSA on the encoding of source information is investigated. An impact of VSA on performance in a source memory test would show that VSA affects not only the MTL/H, but also the PFC component of the Moscovitch model.

Method

Participants. A total of 32 students from the subject pool of the Max-Planck-Institute in Leipzig, who had not participated in experiments 1 to 3, participated. They were paid 3 Euro.

Procedure. The procedure was the same as in experiment 1 and 3 for the study phase and the distractor task, with the exception, that 25 items instead of 26 were presented in each cuing condition. In addition, the time interval between item presentation and presentation of the spatial cue of the next trial was 4 instead of 3 seconds. This was done to assure a source memory performance above chance level. At test, all items of the study phase were presented in random order in the middle of the screen, and the subjects had to indicate the position of the item during encoding. The subjects had to press the 'o'-key (for the German word 'oben') for items that had been presented above fixation, and the 'u'-key (for the German word 'unten') for items that had been presented below fixation. Each test trial had to be initiated by pressing the space bar.

Stimulus material and apparatus. The stimulus material consisted of four lists of 25 German five-letter nouns, which were matched for frequency and concreteness. The assignment of the lists to the study conditions (one list valid, one list invalid) was balanced according to a latin square. The experiment was programmed in *turbopascal.*

Results

Identification performance. Again, significantly more items were identified in the valid cuing condition (.69) than in the invalid cuing condition (.39), $t(31) = 7.1$, $SEM = .043$, $p < .001$, one-tailed.

Source memory performance. Of the items presented in the valid condition, 60 % were correctly assigned to their position during study, while of the invalid condition, 58 % were correctly assigned. This difference was not significant with $t(31) < 1$, $SEM = .027$, $p > .05$, two-tailed. As in the other experiments, fewer items of the invalid condition were identified by the subjects, and we again conditionalized performance upon correct identification in the study phase. Correct source assignments were given to 63 % of identified items of the valid condition and to 67 % of identified items of the invalid study condition. This difference was not significant, $t(31) = 1.3$, $SEM = .027$, $p > .05$, two-tailed. Of the 32 subjects, 14 showed a higher performance for identified valid items, whereas 18 showed a higher performance for identified invalid items.

Discussion

In experiment 4, an impact of VSA on the encoding of source information was found neither for absolute performance nor for performance conditionalized upon correct identification. This result stands in sharp contrast to the large effects of DA on source memory performance and provides first evidence for the notion, that mainly the third component of the Moscovitch model is affected by manipulations of VSA.

CONCLUSION

In four experiments, we investigated the impact of VSA on encoding processes. Our main focus was to find out whether or not VSA influences long-term memory encoding. The second objective of the present research was to examine, to which extent different components of the Moscovitch model are affected, with a focus on the MTL/H and PFC component, and to reveal similarities and differences to the effects of DA on encoding. In all experiments, we presented nouns shortly under valid or invalid cuing conditions. We then applied four different memory tests that are widely accepted as to capture the different memory components. The MTL/H component was represented by a yes-no-recognition test (experiment 1), a cued recall test (experiment 2), and a free recall test (experiment 3). The PFC component was investigated by using a source memory test (experiment 4). Our results are clearcut. In all three tests of the MTL/H component, a significant advantage for items of the valid cuing condition was observed, whether performance was conditionalized upon correct identification or not. VSA seems to affect this memory system. It is interesting to note, that the results of Massen and Stegt (2007), who manipulated VSA in the test phase of a yes-no-recognition task, also suggest that the MTL/H component is affected.

Considering only the recognition results, one might argue that the effects can be explained otherwise. Increased familiarity by improved early perceptual processes could have led to more 'old'-responses for items in the valid cuing condition, because feelings of familiarity, which can be caused by perceptual fluency, also contribute to recognition results. Another objection could be that items that are identified in the valid condition differ from identified items in the invalid condition with respect to word frequency. Identified items in the valid condition could have a lower mean word frequency than corresponding items in the invalid condition. This would favour the former in a recognition test (Glanzer & Adams, 1990).

However, these two explanations are ruled out by the results of experiments 2 and 3. In recall, familiarity contributes less to performance and the results can more easily be attributed to the MTL/H component. In addition, the word frequency effect works in opposite directions for recognition and recall. In recall, it should favour items of the invalid cuing condition. Instead, the results are very similar to those of experiment 1, which strengthens the conclusion that the MTL/H component is affected by VSA. And last but not least, to finally rule out the word frequency explanation of the recognition results, a recent study by Diana and Reder (2006) shows, that under difficult encoding conditions low-frequency words show a larger decrement in recognition performance than do high-frequency words, because they need more processing resources.

Could it be that the advantage of cued items results from a greater allocation of *frontal* resources to these items? The participant might think that these items are more important, provide more executive resources to them and process them more deeply. But if this was the case, one would predict an even more expressed advantage for cued study items in source memory, which was not found in experiment 4. The source memory test revealed no difference between valid and invalid cuing conditions. This is contrary to what might be expected from the effects of DA on source memory versus item memory (Troyer et al., 1999). VSA seems to affect the MTL/H, but not the PFC component of the neuropsychological memory model.

Another alternative explanation is that less time is needed for item identification in the valid condition and therefore more time is available for encoding processes. This hypothesis is ruled out by experiment 2, where a significant advantage for items in the valid condition was found, even though there was an increase (200 ms) of rehearsal time for items in the invalid condition.

The results of the four experiments presented are of theoretical importance for three reasons. First, they show that not only DA but also VSA affects the encoding of items into long-term memory. Second, they provide first evidence for the notion that the impact of VSA on encoding can be dissociated from the impact of DA. DA influences the PFC memory component, which is even more heavily affected than the MTL/H component. Effects of DA during encoding are due to the disruption of organizational and other executive processes, whose product is taken up by the MTL/H component. In contrast, VSA seems to influence the MTL/H, but not the PFC component. Third, from the viewpoint of VSA, as was already suggested by a couple of studies cited above, it emerges that VSA affects not only early perceptual, but also higher-level cognitive processes, that are important for encoding and retrieval and that go beyond the functioning of perceptual modules.

To sum up, the present experiments show, that the impact of attention on memory depends not only on the tests used, but also on the way how attention is manipulated. Further research is needed to investigate the similarities and differences between the effects of DA and VSA on memory. We should have a closer look at the PFC component, and the impact of VSA on performance in tests of implicit memory is also an important issue for further research. In addition, it will be interesting to see, if the impact of IOR on encoding is parallel to that of the classical spatial cuing paradigm used in the present chapter.

REFERENCES

Allport, A. (1993). Attention and control: Have we been asking the wrong questions? A critical review of twenty-five years. In: D. E. Meyer, & S. Kornblum, (Eds.), *Attention and performance XIV.* Synergies in Experimental psychology: Artificial Intelligence and Cognitive Neuroscience (182-218). London: MIT Press.

Alexander, R. & Reinitz, M. T. (2000). Separate and combined effects of location cuing and repetition priming on perceptual encoding of words. *Psychological Science, 11(5),* 419-423.

Baddeley, A. D., Lewis, V., Eldridge, M. & Thomson, N. (1984). Attention and retrieval from long-term memory. *Journal of Experimental Psychology: General, 113,* 518-540.

Bentin, S. Moscovitch, M. & Nirhod, O. (1998). Levels of processing and selective attention effects on encoding in memory. *Acta Psychologica, 98,* 311-341.

Chasteen, A. L. & Pratt, J. (1999). The effect of inhibition of return on lexical access. *Psychological Science, 10,* 41-46.

Craik, F. I. M., Govoni, R., Naveh-Benjamin, M. & Anderson, N. D. (1996). The effects of divided attention on encoding and retrieval processes in human memory. *Journal of Experimental Psychology: General, 125,* 159-180.

Diana, R. A. & Reder, L. M. (2006). The low-frequency encoding disadvantage: word frequency affects processing demands. *Journal of Experimental Psychology: Learning, Memory, and Cognition, 32(4),* 805-815.

Duchek, J. M. & Neely, J. H. (1989). A dissociative word-frequency x levels-of-processing interaction in episodic recognition and lexical decision tasks. *Memory & Cognition, 17(2),* 148-162.

Fernandes, M. A. & Moscovitch, M. (2000). Divided attention and memory: evidence of substantial interference effects at retrieval and encoding. *Journal of Experimental Psychology: General, 129(2),* 155-176.

Fuentes, L. J., Vivas, A. B. & Humphreys, G. W. (1999). Inhibitory tagging of stimulus properties in inhibition of return: Effects on semantic priming and flanker interference. *Quarterly Journal of Experimental Psychology: Human Experimental Psychology, 52,* 149-164.

Gardiner, J. M. & Parkin, A. J. (1990). Attention and recollective experience in recognition memory. *Memory & Cognition, 18,*_579-583.

Glanzer, M. & Adams, J. K. (1990). The mirror effect in recognition memory: Data and theory. *Journal of Experimental Psychology: Learning, Memory, and Cognition, 16,* 5-16.

Glenberg, A. M., Smith, S. M. & Green, C. (1977). Type I rehearsal: Maintenance and more. *Journal of Verbal Learning and Verbal Behavior, 16*, 339-352.

Henson, R. N. A., Hornberger, M. & Rugg, M. D. (2005). Further dissociating the processes involved in recognition memory: An fMRI study. *Journal of Cognitive Neuroscience*, *17*, 1058-1073.

Jacoby, L. L., Toth, J. P. & Yonelinas, A. P. (1993). Separating conscious and unconscious influences of memory: measuring recollection. *Journal of Experimental Psychology, General, 122,* 139-154.

Jacoby, L. L., Woloshyn, V. & Kelley, C. (1989). Becoming famous without being recognized: Unconscious influences of memory produced by divided attention. *Journal of Experimental Psychology: General, 118*, 115-125.

Kelley, C. M. & Jacoby, L. L. (1996). Adult egocentrism: Subjective experience versus analytic bases for judgment. *Journal of Memory and Language, 35*, 157-175.

Krüger, T. (1998). Eine Normierung der Ergänzung der deutschen Wortanfänge zu Substantiven mit fünf oder sechs Buchstaben. *Sprache & Kognition, 17*, 51-72.

Lindsay, D. S. & Kelley, C. M. (1996). Creating illusions of knowing in the remember/know paradigm. *Journal of Memory and Language, 35,* 197-211.

Massen, C. & Stegt, S. J. (2007). Inhibition of return impairs episodic memory access. *Quarterly Journal of Experimental Psychology, 60,* 696-707.

Mayes, A. R. & Montaldi, M. (1999). The neuroimaging of long-term memory encoding processes. *Memory, 7,* 613-660.

Moscovitch, M. (1992). Memory and working-with-memory: a component process model based on modules and central systems. *Journal of Cognitive Neuroscience, 4,* 257-267.

Moscovitch, M. (1994). Cognitive resources and DA interference effects at retrieval in normal people: The role of the frontal lobes and medial temporal cortex. *Neuropsychology, 8,* 524-534.

Mulligan, N. W. (1996). The effects of perceptual interference at encoding on implicit memory, explicit memory, and memory for source. *Journal of Experimental Psychology: Learning, Memory, & Cognition, 22,* 1067-1087.

Mulligan, N. W. (1998). The role of attention during encoding in implicit and explicit memory. *Journal of Experimental Psychology: Learning, Memory, & Cognition, 24,* 27-47.

Nairne, J. S. (1988). The mnemonic value of perceptual identification. *Journal of Experimental Psychology: Learning, Memory, & Cognition, 14,* 248-255.

Naveh-Benjamin, M., Craik, F. I. M., Perretta, J. G. & Tonev, S. (2000). The effects of divided attention on encoding and retrieval processes: the resiliency of retrieval processes. *Quarterly Journal of Experimental Psychology, 53,* 609-626.

Neumann, O. (1992). Theorien der Aufmerksamkeit: Von Metaphern zu Mechanismen. *Psychologische Rundschau, 43(2),* 83-101.

Otten, L. J., Henson, R. N. A. & Rugg, M. D. (2002). State-related and item-related neural correlates of successful memory encoding. *Nature Neuroscience*, *5*, 1339-1344.

Posner, M. I. (1980). Orienting of attention. *Quarterly Journal of Experimental Psychology, 32,* 3-25.

Posner, M. I. & Cohen, Y. (1984). Components of visual orienting. In: H. Bouma, & D. Bouwhuis, (Eds.), *Attention and performance X* (531-556). London: Erlbaum.

Russo, R. & Parkin, A. J. (1993). Age differences in implicit memory: more apparent than real. *Memory & Cognition, 21,* 73-80.

Schneider, W., Eschman, A. & Zuccolotto, A. (2002). *E-Prime user`s guide.* Pittsburgh: Psychology Software Tools Inc.

Snodgrass, J. G. & Corwin, J. (1988). Pragmatics of recognition memory: Applications to dementia and amnesia. *Journal of Experimental Psychology: General, 117*, 34-50.

Troyer, A. K., Winocur, G., Craik, F. I. M. & Moscovitch, M. (1999). Source memory and divided attention: reciprocal costs to primary and secondary tasks. *Neuropsychology, 13,* 467-474.

Tulving, E. (1966). Subjective organization and effects of repetition in multi-trial free-recall learning. *Journal of Verbal Learning and Verbal Behavior, 5*, 193-197.

Tulving, E. (1983). *Elements of episodic memory*. Oxford: University Press.

Tulving, E. (1985). Memory and consciousness. *Canadian Psychologist, 26,* 1-12.

Wheeler, M. E. & Treisman, A. M. (2002). Binding in short-term visual memory. *Journal of Experimental Psychology: General, 131,* 48-64.

Author Note

Stephan Stegt, Institut für Psychologie der Universität Bonn, Römerstr. 164, D-53117 Bonn. Email: Stephan.Stegt@web.de

Cristina Massen, Department of Psychology, Max-Planck-Institute for Human Cognitive and Brain Sciences, Stephanstrasse 1a, 04103 Leipzig, Germany. Email: massen@cbs.mpg.de

This research was funded by German Science Foundation grant MA 2584/1-1 to Cristina Massen and Jürgen Bredenkamp. We thank Anja Heger, Dana Nouzovska, Linda Schüßler, Jürgen Möslein and Martin Ostapczuk for their help with data acquisition.

In: Spatial Memory
Editor: Jacob B. Thomas, pp. 203-226
ISBN: 978-1-61668-139-5

Chapter 9

THE ASSOCIATION OF VISUOSPATIAL MEMORY AND WORKING MEMORY WITH ADOLESCENT ONSET SCHIZOPHRENIA

Alasdair Vance*

Head Academic Child Psychiatry, Department of Paediatrics, University of Melbourne, Murdoch Childrens Research Institute, Royal Children's Hospital, Parkville, Melbourne VIC 3052, Australia

ABSTRACT

This chapter investigates the nature of and association between visuospatial memory, and, separately, visuospatial working memory in adolescent onset schizophrenia. The data presented suggest that both the visuospatial memory and working memory performance of the adolescent-onset schizophrenia group are impaired compared to the healthy control group. Further, the type and magnitude of the deficits are similar to those reported in the adult-onset schizophrenia patients suggesting these deficits are developmental stage independent. The import of these findings is examined within the context of emerging neurobiological models of progressive parietal-temporal-frontal linked neural network dysfunction in schizophrenia, whether of adolescent- or adult-onset. Also, the implications of these findings for the current

* Corresponding author: Phone: 61 3 9345-4666; Fax: 61 3 9345-6002; email: avanc E-@unimelb.edu.au

aetiological understanding of schizophrenia and the current assessment and treatment approaches are discussed.

Introduction

It is known that the cognitive neuroscience constructs of visuospatial memory and working memory are associated with adult- and adolescent-onset schizophrenia. However, to date, the specific associations between well defined measures of visuospatial memory and visuospatial working memory, with known brain behaviour relationships, and carefully defined adolescent onset schizophrenia have not been studied. This chapter reviews the extant literature and explores the nature of and association between visuospatial memory, and, separately, visuospatial working memory in adolescent onset schizophrenia. Data from the cross sectional comparison of adolescent onset schizophrenia with a healthy control group, matched for age, gender and fullscale IQ on the above cognitive neuroscience measures are presented. Both the visuospatial memory and working memory performance of the adolescent-onset schizophrenia group are impaired compared to the healthy control group. Further, the type and magnitude of the deficits are similar to those reported in the adult-onset schizophrenia patients suggesting these deficits are developmental stage independent. The import of these findings is examined within the context of emerging neurobiological models of progressive parietal-temporal-frontal linked neural network dysfunction in schizophrenia, whether of adolescent- or adult-onset. Also, the implications of these findings for the current aetiological understanding of schizophrenia and the current assessment and treatment approaches are discussed.

Study 1

Visuospatial memory deficits are replicated findings in adults (Bilder et al., 2000) and adolescents (Kravariti et al., 2003; Rhinewine et al., 2005) with schizophrenia and those at risk for developing schizophrenia (Brewer et al., 2005; Smith et al., 2006). The magnitude of these deficits ranges between 1-2 standard deviations (Rhinewine et al., 2005) and they are qualitatively similar (Kravariti et al., 2003; Rund et al., 2004). To date, in adults and adolescents with schizophrenia, visuospatial memory deficits have been investigated primarily through the parsing of visuospatial working memory into its maintenance and manipulation components: both of these subprocesses have been found to be deficient in adults (Cannon et al., 2005) and adolescents (Wood et al., 2003; Mathes et al., 2005) with schizophrenia. Also, manipulation abilities are more severely affected than maintenance aspects in visuospatial (Cannon et al., 2005) and verbal (Tan et al., 2005) working memory. Further, the encoding aspect of visuospatial memory, within the broader visuospatial working memory construct, has been specifically implicated (Fleming et al., 1997; Hartman et al., 2002; Liederman and Strejilovich, 2004), while encoding or storage processes have been associated with increasing genetic predisposition to schizophrenia (Glahn et al., 2003).

The delayed matching-to-sample (DMTS) task from the Cambridge Neuropsychological Test Automated Battery (CANTAB) (Owen et al., 1996) is able to readily parse out encoding from retrieval aspects of visuospatial memory for the following reasons: Firstly, the memory demand of the test can be increased by extending the delay between the offset of the target stimulus and the onset of the exemplars. Increasing the delay makes the representation of the encoded stimulus vulnerable to decay, placing greater demand on retrieval processes (Paule et al., 1998). Secondly, by presenting the choice exemplars while the target stimuli remains visible (a matching to sample task (MTS)) the memory demand of the task can be eliminated. The MTS condition provides an ideal control to determine the extent to which factors influencing encoding, such as perceptual discrimination differences, may contribute to later deficits in the recognition or recall of target stimuli (Pantelis et al., 2001). Manipulation of the exemplars such that they share some, but not all, of the target features (for example, the same shape as the target but a different colour) allows the analysis of the types of strategies used by subjects to make decisions based on memory (Owen et al., 1995). Because responses to the DMTS task can be made manually and pictorial stimuli can be used, the extent to which any language impairment alters performance is reduced. Finally, the simultaneous and delayed task conditions are randomized, with all conditions identical until after stimulus presentation (Barnett et al., 2005; Mathes et al., 2005).

However, to date, investigation of DMTS performance in adults with early phase and chronic adult-onset schizophrenia has revealed encoding deficits (Mathes et al., 2005), while encoding and retrieval deficits have been suggested in first episode adult-onset schizophrenia and those at risk for adult-onset schizophrenia (Wood et al., 2003). Given the patients with schizophrenia in the latter study were younger and Luciana et al. (2005) have reported that healthy adolescents aged 16 years and over have significantly better DMTS performance than 13-15 year old healthy adolescents, the encoding and retrieval deficits in the younger patients with schizophrenia may be developmental stage dependent: specifically, the encoding deficits may be contributing to the retrieval deficits in the younger patients rather than there being independent retrieval deficits. This study investigates whether the encoding and/or retrieval phases of visuospatial memory are impaired in carefully defined adolescents with schizophrenia. We hypothesize that that there will be encoding deficits alone, consistent with findings in adults with schizophrenia.

Method

Participants

Nineteen right-handed male adolescents aged 12-17 years [14.64 (1.85) years] were identified with medication naïve, undifferentiated schizophrenia (DSM-IV, American Psychiatric Association, 1994) during their initial clinical evaluation before active treatment (Vance et al., 2007). This was defined through a semi-structured clinical interview (Schedule for Affective Disorders and Schizophrenia – childhood version (K-SADS), Kaufman et al., 1997) with their parent(s) and with each adolescent (κ >0.80) and by the clinician-completed Positive and Negative Symptom Scale (PANSS) (Kay et al., 1987): Positive Scale score of mean 67.11 (23.29) (percentile), a Negative Scale score of 79.53 (20.71) (percentile) and a General Psychopathological Scale score of 89.95 (21.30) (percentile). The undifferentiated schizophrenia group had a fullscale IQ of 109.10 (15.00) (Wechsler Intelligence Scale for Children (WISC-3) (Wechsler, 1991), with all fullscale IQs above 80 and a duration of illness of mean 12.45 (3.67) months. All the group were living in a family home and attending normal schools and none had overt neurological or endocrine disease or colour blindness. Twenty eight healthy male control participants were recruited through local schools, who did not meet criteria for any psychiatric, neurological or endocrine disorder, again defined through a semi-structured clinical interview (K-SADS), Kaufman et al., 1997) with their parent(s) and with each adolescent and by the clinician-completed PANSS [(Kay et al., 1987): Positive Scale score of 2.11 (0.57) (percentile); $t(45) = 14.83$, $p < 0.0005$, a Negative Scale score of 1.71 (0.81) (percentile); $t(45) = 19.97$, $p < 0.0005$ and a General Psychopathological Scale score of 1.04 (0.19) (percentile); $t(45) = 22.20$, $p < 0.0005$] and who did not differ from the undifferentiated schizophrenia group in age [13.47 (1.77); t (45)=1.58, p=0.12) and fullscale IQ [105.89 (6.92); t(45)= 0.98, p=0.33].

Measures

Visuospatial memory was assessed through measurement of performance on the DMTS task selected from the CANTAB (Owen et al., 1996). The computerised test was presented on a high resolution IBM monitor with a touch sensitive screen. All participants reported previous experience with computers. The DMTS test assessed the ability to recall a complex target stimulus (an abstract pattern), presented on the screen for 4 seconds. There were four conditions, each with 10 trials presented randomly. In the MTS condition, the target stimulus remained on the screen with the choice exemplars. In the delay conditions, the target stimulus was removed followed by a 0, 4 or 12 second delay before the choice exemplars were presented. In all trials, subjects were required to choose the target stimulus in a four-alternative forced choice procedure between the correct pattern, and three 'distractor' patterns. Performance was measured according to the percentage of correct responses at the simultaneous and each of the delay levels. Further analysis examined three types of possible errors at each level of the task. 1) a "colour error" was committed when the subject selected the wrong coloured pattern but the correct shape; 2) a "shape error" was committed when the subject selected an incorrectly shaped pattern that was the correct colour, and 3) an "unrelated error" was committed when the incorrect pattern was selected that was both the wrong colour and shape (Barnett et al., 2005).

Statistical Analysis

Latencies for the DMTS task were transformed using logarithmic (base 10) transformations to reduce skewness in the distribution (Maruff et al., 1996). Performance on the DMTS task was compared with a repeated measures multivariate analysis of covariance (MANCOVA) using a two-factor design that included a between-subjects factor (group) and a within subjects factor (delay level), co-varying for the effects of age and fullscale IQ, given the known sensitivity of visuospatial memory measures to these variables (Kravariti et al., 2003; Barnett et al., 2005). To protect against violation of the homogeneity of covariance assumption, analysis of within-group effects or interactions with degrees of freedom greater than one were examined using a repeated measures design within MANCOVA (Wilk's Multivariate test of significance) (Tabachnick and Fidell, 2001). MANCOVA was also performed to analyse the accuracy and latency across delay conditions, covarying for performance on the MTS condition in addition to age and fullscale IQ. To estimate effect sizes, Cohen's d calculation for unequal sample sizes was used (Zakzanis, 2001) to examine differences between group performance on both accuracy and latency measures. In order to protect against a type 1 error, the level of significance for comparisons was set at 0.01. For the number of each type of error made on the DMTS task, the number of each error type made by each subject at each level of the test was submitted to a 2 (group) x 4 (condition) x 3 (error type) repeated measures MANCOVA, with error type as an additional within subject variable, again co-varying for the effects of age and fullscale IQ. Finally, the relationship between response accuracy and response latency, covarying for the effects of age and fullscale IQ, was explored using Pearson product moment correlation coefficients. This same procedure was also used to assess the strength of the relationship between participants' DMTS task performance accuracy, severity of psychotic symptoms (PANSS) and levels of inattention (Child Behaviour Checklist, attention subscale) (Achenbach, 1991), given the known strong association between inattention and memory encoding difficulties (Devilbiss et al., 2006).

Results

Figure 1 shows the group mean percentage correct accuracy of performance and Table 1 the comparison of accuracy rates between the two groups on all conditions of the DMTS test. MANCOVA indicated a significant effect of group [$F(1,43)=18.92$, $p<0.0005$] after controlling for age and full-scale IQ, only the former having a significant independent association with the DMTS accuracy variables [$F(1,43)=28.33$, $p<0.0005$]. When averaged across all four conditions, the control group [88.40 (2.93)] made fewer errors than the undifferentiated schizophrenia group [(80.53 (4.14)]. Figure 2 (filled bar) shows the effect size for this comparison. The MANCOVA also indicated a significant effect of condition [Wilk's $\lambda=0.68$, $F(3,41)=6.36$, $p=0.001$]. There was also a significant group x condition interaction [Wilk's $\lambda=0.75$, $F(3, 41)=4.56$, $p=0.008$], after controlling for age [Wilk's $\lambda=0.69$, $F(3, 41)=6.25$, $p=0.001$], and full-scale IQ [Wilk's $\lambda=0.77$, $F(3, 41)=3.99$, $p=0.01$], both of which had a significant independent association with the DMTS accuracy variables.

To determine the extent to which differences between groups in the accuracy of encoding information affected the accuracy of their recall after a delay, the group x condition MANCOVA was repeated with the three DMTS conditions entered as levels of the independent variable and accuracy on the MTS condition treated as a covariate. This analysis indicated that the covariate was non-significant [$F(1, 42)=1.17$, $p=0.29$], with the effect of group remaining significant when the effect of the covariate was removed [$F(1, 42)=14.10$, $p=0.001$], as did the effect of condition [Wilks $\lambda=0.73$, $F(2, 41)=7.62$, $p=0.002$]. Effect size (Cohen's d) for the corrected difference between conditions is shown in Figure 2 (unfilled bar). This Figure illustrates that the magnitude of the group differences across delay conditions was reduced by approximately 23% with the accuracy of performance on MTS trials removed. The interaction between group and condition became non-significant with the covariate added to the analysis [Wilks $\lambda=0.84$, $F(2, 41)=3.96$, $p=0.03$].

For response latency, MANCOVA indicated a non-significant effect of group [$F(1,43)=1.06$, $p=0.31$] [control group 3.55 (0.10) s; undifferentiated schizophrenia group 3.52 (0.12) s] and condition [Wilk's $\lambda=.84$ $F(3, 41)=2.63$, $p=0.06$]. After controlling for age and fullscale IQ, both of which had no significant independent association with DMTS latency variables, there was a significant group x condition interaction [Wilk's $\lambda=0.66$, $F(3, 41)=6.95$, $p=0.001$].

To determine the extent to which differences between groups in the speed of encoding information affected the speed of their recall after a delay, the group x condition MANCOVA was repeated with the log 10 latencies for each subject as the dependent variable and the three DMTS conditions entered as levels of the repeated factor and log 10 latencies on the MTS condition treated as a covariate. This analysis indicated that the covariate was significant [$F(1, 42)=23.15$, $p<0.0005$]. The effect of group became significant with the effect of the covariate removed [$F(1, 42)=7.34$, $p<0.01$]. The effect of condition remained non-significant with the covariate removed [Wilks $\lambda=0.95$, $F(2, 41)=1.15$, $p=0.33$], although, as Figure 2 shows, the effect size for the difference between groups altered substantially with the covariate removed (unfilled bar). The interaction between group and condition remained significant with the covariate added to the analysis [Wilks $\lambda=0.70$, $F(2, 41)=8.85$, $p=0.001$], after controlling for age and full-scale IQ, both of which had no significant independent association with DMTS latency variables.

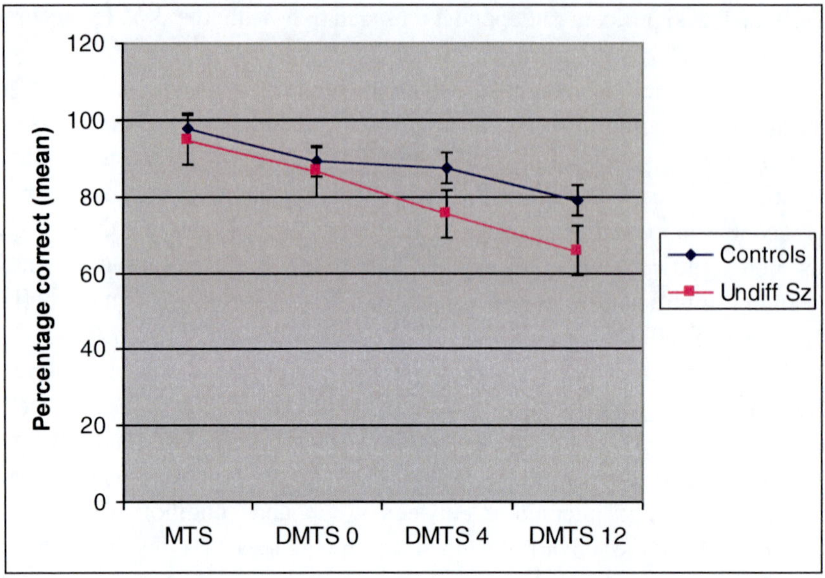

Figure 1. Response accuracy of each group at each condition of the DMTS task.

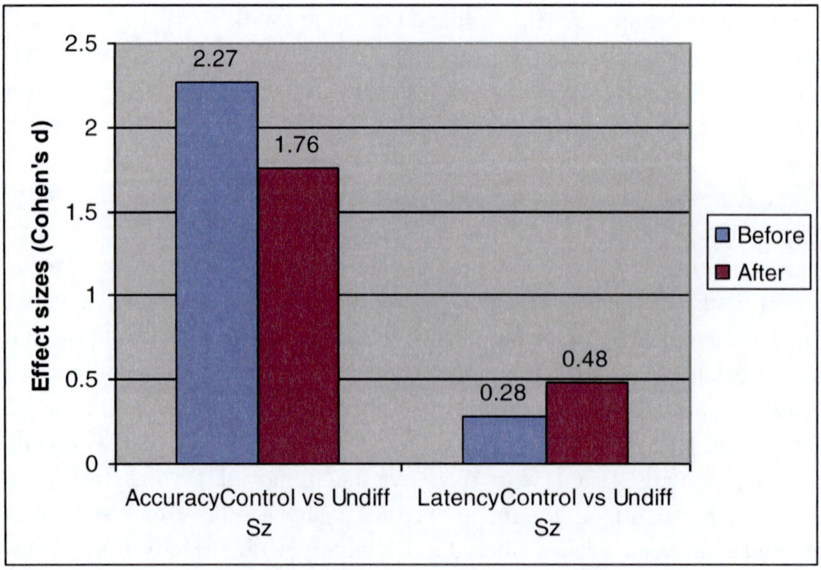

Figure 2. Effect sizes before and after covarying for performance on the MTS condition.

For analysis of error type, MANCOVA indicated a significant main effect of group [F(1,42)=18.09, p <0.0005]. As the differences between groups for the mean number of errors, averaged across all error types, is the inverse of the finding of differences between groups for accuracy of performance (shown above), this effect was not explored further. MANCOVA also indicated a significant main effect for condition [Wilk's λ=0.63, F(3,40)=7.56, p <0.0005], while error type and condition x error type interaction were non-significant.

For the MTS condition, no significant correlations were found between response accuracy and Log10 latency of correct responses in the control [r=0.22, p =0.27] or undifferentiated schizophrenia [r=-0.03, p =0.90] groups. No significant correlation was found between accuracy on the delay conditions (percentage correct averaged across 0, 4 and 12 second delay conditions) and latency for the control [r= 0.15, p =0.46] or undifferentiated schizophrenia [r=0.40, p =0.09] groups.

No significant correlations were found between DMTS task performance and positive [r=-0.30, p =0.16] or negative [r=-0.45, p =0.11] psychotic symptom severity or levels of inattention [r=-0.28, p =0.27] in the undifferentiated schizophrenia group. Similarly, no significant correlations were found between DMTS task performance and positive [r=-0.24, p =0.25] or negative[r=-0.16, p =0.42] psychotic symptom severity or levels of inattention [r=-0.22, p =0.27] in the control group.

Table 1. Comparison of accuracy rates across each condition of the delayed-matching-to sample test

	Controls N = 28 Mean (SD)	Sz N = 19 Mean (SD)	p
MTS trials (% correct)	97.86 (6.30)	94.74 (8.41)	0.15
DMTS 0 second delay (% correct)	89.29 (13.03)	86.32 (13.00)	0.45
DMTS 4 second delay (% correct)	87.50 (12.95)	75.26 (20.65)	0.02
DMTS12 second delay (% correct)	78.93 (14.49)	65.79 (24.11)	0.02
DMTS delay trials (% correct averaged across delay trials only)	85.24 (11.84)	75.79 (17.24)	0.03

Controls: healthy control participants. Sz: undifferentiated schizophrenia participants.

Discussion

Anti-psychotic medication naïve adolescents with undifferentiated schizophrenia demonstrated subtle yet significant impairment in performance on a visuospatial memory test. Although performance on the memory test was generally worse in the undifferentiated schizophrenia group than the healthy control participants, the pattern of decrease in overall accuracy from the MTS to the 12 second DMTS condition was qualitatively similar in both groups (Figure 1). Further, although accuracy decreased substantially from the MTS to DMTS conditions for both groups, they could not be differentiated statistically across the 0 to 4 second and from the 4 to 12 second delay conditions. When the 'visible exemplar' condition (MTS) and the 'removed from view' exemplar conditions (DMTS) were examined together, an interaction effect was obtained suggesting that the undifferentiated schizophrenia group behaved differently to the healthy control participants across these stages. However, this interaction effect disappeared when DMTS performance between the two groups was examined covarying for the MTS performance. Taken together, these findings suggest that the undifferentiated schizophrenia group are generally worse at remembering the exemplars but not at any particular delay. Also, although the medication naïve adolescents with undifferentiated schizophrenia were significantly faster at responding on the delay conditions than the healthy control participants, there was no correlation between the speed of response and accuracy of DMTS task performance in either group. Further, error type did not differ between the two groups on the MTS and delay conditions. Hence, the difference between DMTS task performance between the two groups is unlikely to be due to motivational differences and/or manifestly different approaches to task completion. In particular, the adolescents with undifferentiated schizophrenia appear to be using the same decision making strategy, obeying the task rules, adhering to the task requirements and not selecting targets at random (Barnett et al., 2005). Similarly, the absence of any correlation between the level of psychotic symptoms, level of inattention and DMTS task performance in either group suggests that the DMTS performance impairments in the undifferentiated schizophrenia group are not secondary to these behavioural symptoms.

These findings are consistent with the adolescents with undifferentiated schizophrenia experiencing an encoding impairment rather than a problem with retrieval (Barnett et al., 2005). Consequently, poorer performance during the delay conditions of the task may have occurred because the adolescents with undifferentiated schizophrenia had not encoded the "to-be-remembered" stimulus effectively when it was initially shown, rather than forgetting what they had learned. Memory impairment did not increase with increasing delay and it is a decline in memory accuracy as a function of increasing delay (Sahakian et al., 1988) that is consistent with findings in patients with amnesia and human and non-human primates with lesions of the hippocampal formation, amygdala or medial temporal lobes (Sahakian et al., 1988; Joyce and Robbins, 1993; Owen et al., 1995; Elliott and Dolan, 1999; Chelonis et al., 2000).

Our findings add to the two extant studies in adults at risk for adult-onset schizophrenia or with early phase adult-onset schizophrenia (Wood et al., 2003; Mathes et al., 2005) and Luciana et al.'s (2005) developmental study of DMTS performance by suggesting that adolescents with undifferentiated schizophrenia have encoding but not retrieval deficits. Further, the sample studied addressed the potential neurodevelopmental confounds of gender, IQ and handedness.

Currently, it is thought that disruption to fronto-striatal dopaminergic neurocognitive systems interferes primarily with the encoding of stimuli and this can lead to memory impairments when those stimuli must be recognised after a delay. However, this impairment is independent of the length of time for which information must be held in memory. Specifically, activation of the caudate nucleus extending to the subgenual cingulate cortex (BA 24), the medial orbitofrontal cortex (BA 25 extending to 11) and the superior parietal cortex (BA 7) is seen during shorter delays on the DMTS task (Elliot and Dolan, 1999). In contrast, the published data (Sahakian et al., 1988; Lange et al., 1992; Robbins et al., 1997; Lawrence et al., 2000) suggest that delay-dependent memory impairments on the DMTS reflect disruption to cholinergically modulated neural circuits within the temporal lobe. Specifically, activation of the temporal pole (BA 21/38), medial temporal gyrus (BA 21), inferior temporal gyrus (BA 20), hippocampus and inferior frontal gyrus (BA 44/45) are associated with longer delays on the DMTS task (Elliot and Dolan, 1999) while the ability to encode stimuli is not disrupted. Our current results suggest only encoding fronto-striatal but not retrieval fronto-temporal neural systems are impaired in the medication naïve adolescents with undifferentiated schizophrenia. This finding is consistent with a developmental stage-independent dysfunction in schizophrenia. Importantly, these findings are consonant with the current conceptualization of schizophrenia as a neurodevelopmental disorder involving progressive gray and white matter changes from the parietal - temporal - frontal lobes (Thompson et al., 2001; Vidal et al., 2006).

In summary, medication naïve adolescents with undifferentiated schizophrenia were clearly only impaired in their encoding and accurately selecting the pattern of stimuli on the visuospatial memory task. Of clinical importance, adolescents with undifferentiated schizophrenia may demonstrate memory impairments when performing everyday and academically-related tasks. The current results suggest that such problems may be best dealt with by encouraging encoding of relevant information, rather than emphasising retrieval or recall strategies (that is, memory lists): Careful study of adolescent undifferentiated schizophrenia samples does detect deficits in a DMTS memory task, but these reflect attentional deficits rather than being specifically due to fronto-temporally mediated explicit memory system dysfunction.

The next study will address the nature of and association between visuospatial working memory and adolescent onset schizophrenia.

STUDY 2

Dysfunctional working memory systems subserved by the prefrontal cortex (PFC) (Goldman-Rakic and Selemon, 1997) may be associated with the core negative and disorganized symptoms of schizophrenia. To date, visuospatial working memory (VSWM) has been investigated because this construct is (1) largely non-verbal, (2) models of the relationship between VSWM and PFC are constrained by single neuronal firing rates in alert primates (Goldman –Rakic, 1987), studies of behaviour following focal PFC lesions in adult humans (Owen et al., 1993) and studies of patterns of regional cerebral blood flow in healthy humans performing VSWM tests (D'Esposito et al., 1995) and (3) the experimental paradigm used in these studies can be applied unchanged to the study of VSWM in children and adolescents (Barnett et al., 2001). Further the Cambridge Neuropsychological Test Automated Battery (CANTAB) (Owen et al., 1996) VSWM task has the additional benefit of allowing two key associated components of the dependent VSWM measure, 'between search error' (BSE) performance to be measured: (1) the ability to develop a systematic strategy during task performance (as reflected in the 'spatial strategy' measure) and (2) the ability to hold information 'on line' in a visuospatial scratchpad (Baddeley, 1986) (as reflected in the 'spatial span' measure) (Owen et al., 1990; Robbins, 1996; Panetlis et al., 1997). Hence, VSWM deficits, suggested by poor BSE performance, can be further analysed to see if this poor BSE performance relates to either or both of these components (Pantelis et al., 1997; Barnett et al., 2001).

This approach of parsing the construct of VSWM has been extended to the examination of putative underlying neural networks that may subserve the construct of VSWM and these key components. VSWM (as defined by 'BSE' or comparable measure performance) has been associated with activation of frontal-striatal-parietal neural networks, particularly the PFC and posterior parietal brain regions (Carlson et al., 1998; Braver et al., 1997; 2001). With respect to the PFC, VSWM is thought to 'rely' on the integrity of the dorsolateral prefrontal cortex (DLPFC) in non-human primates and human beings (Goldman-Rakic and Selemon, 1997; Cannon et al., 2005). Further, Owen et al. (1996) investigated VSWM in a PET study using tasks derived from 'BSE', 'spatial strategy' and 'spatial span' CANTAB measures. They reported that the 'organising, executing and monitoring' component, involving active monitoring and manipulation of spatial information (as reflected in 'spatial strategy'), may be associated with the mid-DLPFC (BA 46, 9), while the 'spatial span' component, involving a sequence of retained spatial moves in working memory, may be subserved by the ventrolateral prefrontal cortex (Brodmann Area (BA) 47).

To date, VSWM deficits (as defined by 'BSE' or comparable measure performance) have been shown in the acute and chronic phases of adult-onset schizophrenia (Pantelis et al., 1997; Wood et al., 2003) and those 'at-risk' for adult-onset schizophrenia (Glahn et al., 2003; Wood et al., 2003). Also, VSWM deficits are qualitatively similar but less severe in those with a broader phenotype of schizophrenia (Tsuang et al., 2002). Specifically, these VSWM deficits are associated with the 'spatial span' component in chronic adult-onset schizophrenia (Pantelis et al., 1997), acute adult-onset schizophrenia and those 'at-risk' for adult-onset schizophrenia (Wood et al., 2003), evident through between groups and correlational analyses. This study investigates whether VSWM deficits are (1) evident in adolescent-onset schizophrenia and (2) whether the relationships between components of VSWM are similar in adolescents to those reported in adults with schizophrenia.

Method

Participants

Nineteen right-handed male adolescents aged 12-17 years [14.64 (1.85) years] were identified with medication naïve, undifferentiated schizophrenia

(DSM-IV, American Psychiatric Association, 1994) during their initial clinical evaluation before active treatment (Vance et al., 2006). This was defined through a semi-structured clinical interview (Schedule for Affective Disorders and Schizophrenia – childhood version (K-SADS), Kaufman et al., 1997) with their parent(s) and with each adolescent [κ >0.80] and by the clinician-completed Positive and Negative Symptom Scale (PANSS) (Kay et al., 1987): Positive Scale score of 67.11 (23.29) (percentile), a Negative Scale score of 79.53 (20.71) (percentile) and a General Psychopathological Scale score of 89.95 (21.30) (percentile). The undifferentiated schizophrenia group had a fullscale IQ of 109.10 (15.00) (Wechsler Intelligence Scale for Children (WISC-3) (Wechsler, 1991), with all fullscale IQs above 70 and a duration of illness of 12.45 (3.67) months. All the group were living in a family home and attending normal schools and none had overt neurological or endocrine disease or colour blindness. Twenty eight healthy male control participants were recruited through local schools, who did not meet criteria for any psychiatric, neurological or endocrine disorder, again defined through a semi-structured clinical interview (K-SADS), Kaufman et al., 1997) with their parent(s) and with each adolescent and by the clinician-completed PANSS [(Kay et al.,1987): Positive Scale score of 2.11 (0.57) (percentile); $t(45) = 14.83$, $p < 0.0005$, a Negative Scale score of 1.71 (0.81) (percentile); $t(45) = 19.97$, $p < 0.0005$ and a General Psychopathological Scale score of 1.04 (0.19) (percentile); $t(45) = 22.20$, $p < 0.0005$] and who did not differ from the undifferentiated schizophrenia group in age [13.47 (1.77); $t(45)=1.58$, $p=0.12$] and fullscale IQ [105.89 (6.92); $t(45)= 0.98$, $p=0.33$].

Measures

Visuospatial working memory was assessed through measurement of performance on the Spatial Working Memory and Spatial Span tasks selected from the CANTAB (Owen et al., 1996). The computerised tests were presented on a high resolution IBM monitor with a touch sensitive screen. The Spatial Working Memory task is a self-ordered searching task that measures working memory for spatial stimuli and requires the subject to use mnemonic information to work towards a goal. Subjects are required to search through boxes that appear on the screen with the aim of finding the 'blue tokens' hidden inside. The key instruction was that once a token had been taken out of a box, that box would not be used again to hide a token. After two practice trials with two boxes, there were four test trials with each of two, three, four, six and eight boxes. Returning to an 'empty' box already opened and a token removed on a previous search constituted a 'forgetting' or 'Between Search' error (BSE). A Strategy score was calculated from subject's performance on the six and eight box levels, to reflect how often a searching sequence was initiated from the same box during a given trial. Higher Strategy scores represent low use of strategy (i.e., many sequences beginning with a different box in a given trial), and lower scores represent efficient use of strategy (i.e., many sequences starting with the same box in a given trial). The Spatial Span task is a computerised version of the Corsi block tapping test that assesses spatial short term memory capacity. This task measures the ability to remember a sequence of squares presented on the screen. After an incorrect attempt at choosing the squares in sequence, the next trial remains at the same difficulty level. The Spatial Span was calculated at the highest level at which the subject successfully remembered at least one sequence of boxes.

Statistical Analysis

Performance on the (1) BSE variable and (2) the Strategy and Spatial Span variables was compared between groups using respective (1) repeated measures and (2) one-way analysis of covariance (ANCOVA), co-varying for the effects of age and fullscale IQ, given the known sensitivity of these VSWM measures to these variables (Pantelis et al., 1997). To protect against experiment-wise error Type I error rates were set at $p<0.01$. To investigate the contribution of Spatial Span and Strategy to BSE, the intercorrelations between BSE and Span and between BSE and Strategy for all levels of difficulty were investigated for each group using Pearson product moment correlation coefficients.

Results

In the spatial working memory task, there was a significant effect of condition (number of boxes) [Wilks' $\lambda = 0.31$, $F(4,40) = 21.46$, $p < 0.0005$] and group [$F(1,43) = 9.54$, $p = 0.004$, partial $\kappa 2 = 0.19$] with the undifferentiated schizophrenia group being associated with increased BSE, after controlling for age and fullscale IQ, only the latter having a significant independent association with BSE [$F(1,43) = 25.82$, $p < 0.0005$, partial $\kappa 2$ =

0.38] (Figure 1). Also, there was a significant interaction between BSE and group [Wilks' $\lambda = 0.66$, $F(4,40) = 4.99$, $p = 0.002$], after controlling for age and fullscale IQ, again only the latter having a significant independent association with BSE [fullscale IQ: Wilks' $\lambda = 0.43$, $F(4,40) = 12.99$, $p < 0.0005$]. Neither Spatial Span [$F(1,43) = 0.91$, $p = 0.35$, partial $\kappa 2 = 0.02$] (undifferentiated schizophrenia, 6.06 (1.21); control 6.39 (1.37)) nor Strategy [$F(1,43) = 0.42$, $p = 0.52$, partial $\kappa 2 = 0.01$] (undifferentiated schizophrenia, 31.99 (6.78); control 31.75 (3.97)) differed between the groups, after controlling for age and fullscale IQ, which had no independent association with either.

The pattern of intercorrelations between BSE, Spatial Span and Strategy was qualitatively different for the control [BSE: Spatial Span $r = -0.11$, $p = 0.59$; BSE: Strategy $r = 0.31$, $p = 0.12$] and undifferentiated schizophrenia groups [BSE: Spatial Span $r = -0.26$, $p = 0.32$; BSE: Strategy $r = 0.51$, $p = 0.04$], after controlling for age and fullscale IQ. For both groups, there was no correlation between Spatial Span and BSE. However there was a medium-large correlation between Strategy and BSE in the undifferentiated schizophrenia alone.

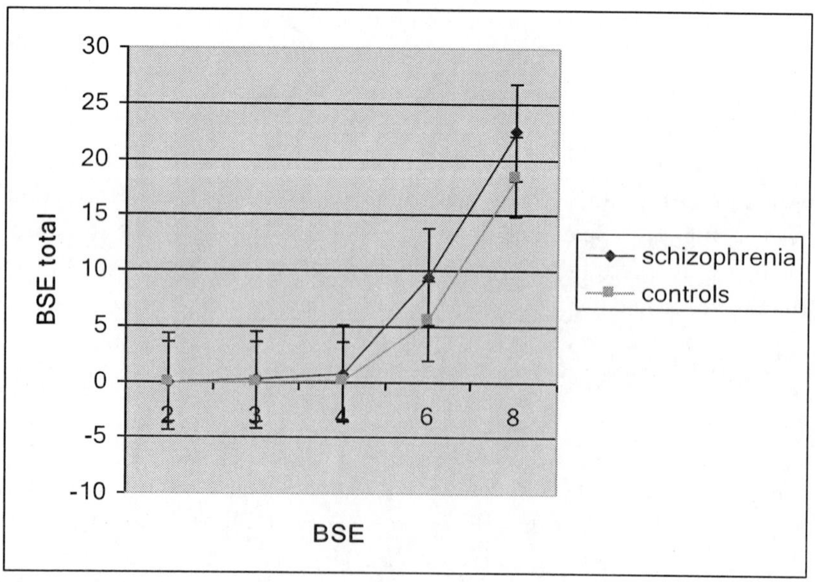

Figure 3. Between Search Errors (BSE - total score with standard error bars) at the 2,3,4,6 and 8 box levels of difficulty in the undifferentiated schizophrenia and healthy control groups.

Discussion

The adolescent-onset schizophrenia group is associated with increased negative and positive symptoms, consistent with previous studies of schizophrenia in this developmental phase (Schultz et al., 2000; Hollis et al., 2003). VSWM deficits (as defined by poor 'BSE' performance) are also evident, commensurate with Goldman-Rakic and Selemon's (1997) model and findings in adult-onset schizophrenia and the broader schizophrenia phenotype (Wood et al., 2003). Further, spatial span is spared in the adolescent-onset schizophrenia group in contrast to comparable data in adult-onset schizophrenia groups during 'at-risk', acute and chronic phases of illness (Pantelis et al., 1997; Wood et al., 2003), evident through between groups and correlational analyses. Spatial strategy is also spared in the adolescent-onset schizophrenia group consistent with comparable data in adult-onset schizophrenia groups (Wood et al., 2003), evident through between groups analysis. However, correlational analysis suggests that VSWM deficits (as defined by poor 'BSE' performance) are associated with spatial strategy in the adolescent-onset group unlike the adult-onset schizophrenia groups during 'at-risk', acute and chronic phases of illness (Pantelis et al., 1997; Wood et al., 2003).

These findings imply that frontal-striatal-parietal neural networks are dysfunctional in adolescent-onset schizophrenia, while mid-dorsolateral and ventrolateral PFC functions remain intact. In addition, the adolescent-onset schizophrenia group relied on spatial strategy rather than spatial span in their overall VSWM performance. In light of Wood et al.'s (2003) findings, this may suggest that the mid-dorsolateral PFC-linked fronto-parietal systems may be relatively more robust than the ventrolateral PFC fronto-parietal neural networks: there is emerging evidence that the ventrolateral PFC fronto-parietal neural networks (as reflected in the 'spatial span' component of VSWM) are less developed in adolescents compared to adults, whether schizophrenia is present or not (Saper et al., 2000; Luciana et al., 2005). As these fronto-parietal systems develop into adulthood, deficits in the ventrolateral PFC fronto-parietal neural networks (as reflected in the 'spatial span' component of VSWM) become apparent in adult-onset schizophrenia and those with the broader phenotype.

These possible progressive changes from one developmental phase to another are consistent with the current conceptualisation of schizophrenia as a neurodevelopmental disorder with progressive gray and white matter changes affecting the parietal through to frontal lobes (Thompson et al., 2001; Gogtay et al., 2004; Vidal et al., 2006). However, this study does not address the specificity of VSWM deficits for schizophrenia: future studies need to compare VSWM and its components performance across developmental phases in those with schizophrenia, those at risk of schizophrenia and those with the broader phenotype.

INTEGRATION AND IMPLICATIONS OF THESE FINDINGS

The findings from both of these studies support the neurodevelopmental nature of schizophrenia, regardless of its age of onset. In particular, the findings are congruent with the known progressive maturation of gray and white matter neural networks in a caudal to rostral direction; that is from parietal through temporal to frontal regions. For instance, the determination of visuospatial memory encoding deficits in adolescents with undifferentiated schizophrenia is consistent with fronto-striatal-parietal neural network dysfunction that may arise from mainly parietal lobe region deficits. Visuospatial memory retrieval deficits, subserved by fronto-striatal-temporal neural networks, were not evident in the adolescents, unlike adults with schizophrenia. Hence, temporal lobe region-linked deficits may not yet be evident in the adolescents as they are in adults with schizophrenia. Similarly, visuospatial working memory deficits were evident in the adolescents with undifferentiated schizophrenia, implicating fronto-striatal-parietal neural network dysfunction while mid-dorsolateral and ventrolateral prefrontal cortical functions remained intact.

Future clinical research needs to [1] apply multiple cognitive neuroscience measures with well defined brain behaviour relationships that are able to parse visuospatial memory into encoding and retrieval components and working memory into capacitance and strategising components. Visuospatial rather than verbal memory and working memory measures remain preferable because they are relatively free of human language. Therefore, language learning and/or proficiency difficulties or disorders will not confound their interpretation. In addition, future clinical research needs to [2] apply functional magnetic resonance imaging (fMRI) paradigms that arise from these specific and targeted cognitive neuroscience measures; and investigate the associations of [1] and [2] with key candidate gene polymorphisms to systematically elucidate the neurodevelopmental pathophysiology of schizophrenia across key developmental stages such as adolescence and adulthood. Also, different subtypes of schizophrenia need to be systematically explored, specifically, disorganized, catatonic and paranoid forms of schizophrenia across key cultural and racial groups in which genetic influences may vary.

From a clinical perspective, these findings emphasize the importance of thorough clinical neuropsychological assessment of patients with schizophrenia. This assessment enables the inpatient and outpatient environments and treatment plan to be tailored to each given patient's executive functioning ability. Further, adolescents with schizophrenia require specific aid to encode information while adults require this help along with memory aids for their retrieval abilities. Possible future randomized controlled trials of stimulant medication, atomoxetine, and/or third line treatments for Attention Deficit Hyperactivity Disorder such as clonidine, imipramine or risperidone may be useful to best treat these memory encoding and working memory deficits.

REFERENCES

Achenbach, T. M. (1991). *Manual for the Child Behavior Checklist* / 4-18 and 1991 Profile. University of Vermont, Burlington.

American Psychiatric Association, (1994). Diagnostic and Statistical Manual of Mental Disorders, fourth edition. *American Psychiatric*, Press, Washington, DC.

Baddeley, A. D. (1986). *Working Memory*. Clarendon Press, Oxford.

Barnett, R., Vance, A. L. A., Maruff, P., Luk, S. L., Costin, J. & Pantelis, C. (2001). Abnormal executive function in attention deficit hyperactivity disorder: The effect of stimulant medication and age on spatial working memory. *Psychological Medicine*, *31*, 1107-1115.

Barnett, R., Maruff, P. & Vance, A. (2005). An investigation of visuospatial memory impairment in children with ADHD, combined type. *Psychological Medicine*, *35*, 1433-1443.

Bilder, R., Goldman, R. & Robinson, D. (2000). Neuropsychology of first-episode schizophrenia: initial characterization and clinical correlates. *American Journal of Psychiatry*, *157*, 549-559.

Braver, T. S., Cohen, J. D., Nystrom, L. E., Jonides, J., Smith, E. E. & Noll, D. C. (1997). A parametric study of prefrontal cortex involvement in human working memory. *Neuroimage*, *5*, 49-62.

Braver, T. S., Barch, D. M., Kelley, W. M., Buckner, R. L., Cohen, N. J., Miezen, F. M., Snyder, A. Z., Ollinger, J. M., Akbudak, E., Conturo, T. E. & Petersen, S. E. (2001). Direct comparison of prefrontal cortex regions engaged by working and long-term memory tasks. *Neuroimage*, *14*, 48-59.

Brewer, W., Francey, S. & Wood, S. (2005). Memory impairments identified in people at ultra-high risk for psychosis who later develop first-episode psychosis. *American Journal of Psychiatry*, *162*, 71-78.

Cannon, T. D., Glahn, D. C., Kim, J., Van Erp, T., Karlsgodt, K., Cohen, M. S., Nuechterlein, K. H., Bava, S. & Shirinyan, D. (2005). Dorsolateral prefrontal cortex activity during maintenance and manipulation of information in working memory in patients with schizophrenia. *Archives of General Psychiatry*, *62*, 1071-1080.

Carlson, S., Martinkauppi, S., Rama, P., Salli, E., Korvenoja, A. & Aronen, H. J. (1998). Distribution of cortical activation during visuospatial n-back tasks as revealed by functional magnetic resonance imaging. *Cerebral Cortex*, *8*, 743-752.

Chelonis, J. J., Daniels-Shaw, J. L. & Blake, D. J. (2000). Developmental aspects of delayed matching-to-sample task performance in children. *Neurotoxicology Teratology*, *22*, 683-694.

D'Esposito, M., Detre, J. A., Alsop, D. C., Shin, R. K., Atlas, S. & Grossman, M. (1995). The neural basis of the central executive system of working memory. *Nature*, *378*, 279-281.

Devilbiss, D. M., Page, M. E. & Waterhouse, B. D. (2006). Locus ceruleus regulates sensory encoding by neurons and networks in waking animals. *Journal of Neuroscience*, *26*, 9860-9872.

Elliot, R. & Dolan, R. J. (1999). Differential neural responses during performance of matching and non-matching to sample tasks at two delay conditions. *Journal of Neuroscience*, *19*, 5066-5073.

Fleming, K., Goldberg, T. E., Binks, S., Randolph, C., Gold, J. M. & Weinberger, D. R. (1997). Visuospatial working memory in patients with schizophrenia. *Biological Psychiatry*, *41*, 43-49.

Giedd, J. N., Lenane, M., Gochman, P., Evans, A. & Rapoport, J. L. (2004). Comparison of progressive cortical gray matter loss in childhood-onset schizophrenia with that in childhood-onset atypical psychoses. *Archives of General Psychiatry*, *61*, 17-22.

Glahn, D. C., Therman, S., Manninen, M., Huttunen, M., Kaprio, J., Lonnqvist, J. & Cannon, T. D. (2003). Spatial working memory as an endophenotype of schizophrenia. *Biological Psychiatry*, *53*, 624-626.

Gogtay, N., Sporn, A., Clasen, L. S., Nugent, T. F., Greenstein, D., Nicolson, R. & Goldman-Rakic, P. S. (1987). Circuitry of the primate prefrontal cortex and the regulation of behaviour by representational memory. In: F., Plum, (Eds.), *Handbook of Physiology, The Nervous System, Higher functions of the Brain. Vol. V*. American Physiological Society, Maryland, 373-417.

Goldman-Rakic, P. S. & Selemon, L. D. (1997). Functional and anatomical aspects of prefrontal pathology in schizophrenia. *Schizophrenia Bulletin*, *23*, 437-458.

Hartman, M., Steketee, M., Silva, S., Lanning, K. & McCann, H. (2002). Working memory and schizopohrenia: evidence for slowed encoding. *Schizophrenia Research*, *59*, 99-113.

Hollis, C. (2003). Developmental precursors of child- and adolescent-onset schizophrenia and affective psychoses: diagnostic specificity and continuity with symptom dimensions. *British Journal of Psychiatry*, *182*, 37-44.

Joyce, E. M. & Robbins, T. W. (1991). Frontal lobe function in Korsakoff and non-Korsakoff alcoholics: planning and spatial working memory. *Neuropsychologica*, *29*, 709-723.

Kaufman, J., Birmaher, B. & Brent, D. (1997). Schedule for Affective Disorders and Schizophrenia – present and lifetime version (K-SADS-PL): initial reliability and validity data. *Journal of the American Academy of Child and Adolescent Psychiatry*, *36*, 980-988.

Kay, S. R., Fiszbein, A. & Opler, L. A. (1987). The Positive and Negative Syndrome Scale (PANSS) for schizophrenia. *Schizophrenia Bulletin*, *13*, 261-276.

Kravariti, E., Morris, R., Rabe-Hesketh, S., Murray, R. M. & Frangou, S. (2003). The Maudsley Early-Onset Schizophrenia study: cognitive function in recent onset schizophrenia. *Schizophrenia Research*, *61*, 137-148.

Kravariti, E., Morris, R., Rabe-Hesketh, S., Murray, R. M. & Frangou, S. (2003). The Maudsley Early-Onset Schizophrenia study: cognitive function in adolescent-onset schizophrenia. *Schizophrenia Research*, *65*, 95-103.

Lange, K. W., Robbins, T. W. & Marsden, C. D. (1992). L-Dopa withdrawal in Parkinson's disease selectively impairs cognitive performance in tests sensitive to frontal dysfunction. *Psychopharmacology*, *107*, 394-404.

Lawrence, A. D., Watkins, L. H. A. & Sahakian, B. J. (2000). Visual object and visuospatial cognition in Huntington's disease: implications for information processing in corticostriatal circuits. *Brain*, *123*, 1349-1364.

Leiderman, E. & Strejilevich, S. (2004). Visuospatial deficits in schizophrenia: central executive and memory subsystem impairments. *Schizophrenia Research*, *68*, 217-223.

Luciana, M., Conklin, H. M., Hooper, C. J. & Yarger, R. S. (2005). The development of nonverbal working memory and executive control processes in adolescents. *Child Development*, *76*, 697-712.

Maruff, P., Tyler, P. & Burt, T. (1996). Cognitive deficits in Machado-Joseph Disease. *Annals of Neurology*, *40*, 421-427.

Mathes, B., Wood, S. J. & Proffitt, T. M. (2005). Early processing deficits in object working memory in first episode schizophreniform psychosis and established schizophrenia. *Psychological Medicine*, *35*, 1053-1062.

Owen, A. M., Downes, J. J., Sahakian, B. J., Polkey, C. E. & Robbins, T. W. (1990). Planning and spatial working memory following frontal lesions in man. *Neuropsychologia*, *28*, 1021-1034.

Owen, A. M., Roberts, A. C., Hodges, J. R., Summers, B. A., Polkey, C. E. & Robbins, T. W. (1993). Contrasting mechanisms of attentional set-shiftng in patients with frontal lobe damage or Parkinson's Disease. *Brain*, *116*, 1159-1175.

Owen, A. M., Sahakian, B. J. & Semple, J. (1995). Visuo-spatial short-term recognition memory and learning after temporal lobe excisions, frontal lobe excisions or amygdalo-hippocampectomy in man. *Neuropsychologia*, *33*, 1-24.

Owen, A. M., Evans, A. C. & Petrides, M. (1996). Evidence for a two-stage model of spatial working memory processing within the lateral frontal cortex: a positron emission tomography study. *Cerebral Cortex*, *6*, 31-38.

Pantelis, C., Barnes, T. R. E., Nelson, H. E., Tanner, S., Weatherley, L., Owen, A. M. & Robbins, T. W. (1997). Frontal-striatal cognitive deficits in patients with chronic schizophrenia. *Brain*, *120*, 1823-1843.

Pantelis, C., Stuart, G. & Hazel, N. (2001). Spatial working memory deficits in schizophrenia: relationship with tardive dyskinesia and negative symptoms. *American Journal of Psychiatry*, *158*, 1276-1285.

Paule, M. G., Bushnell, P. J. & Maurissen, J. P. J. (1998). Symposium Overview: The use of delayed matching-to-sample procedures in studies of short-term memory in animals and humans. *Neurotoxicology Teratology*, *20*, 493-502.

Rhinewine, J., Lencz, T. & Thaden, E. P. (2005). Neurocognitive profile in adolescents with early onset schizophrenia: clinical correlates. *Biological Psychiatry*, *58*, 705-712.

Robbins, T. W. (1996). Dissociating executive functions of the prefrontal cortex. *Philosophical Transactions of the Royal Society of London B Biological Sciences*, *351*, 1463-1470.

Robbins, T .W., Semple, J. & Kumar, R. (1997). Effects of scopolamine on delayed matching to sample and paired associates of visual memory and learning in human subjects: comparison with diazepam and implications for dementia. *Psychopharmacology 134*, 95-106.

Rund, B. R., Melle, I., Friis, S., Larson, T. K., Midboe, L. J., Opjordsmeon, S., Simonsen, E., Vaglum, P. & McGlashan, T. (2004). Neurocognitive dysfunction in first-episode psychosis: correlates with symptoms, premorbid adjustment and duration of untreated psychosis. *American Journal of Psychiatry*, *161*, 466-472.

Sahakian, B. J., Morris, R. G. & Evenden, J. L. (1988). A comparative study of visuospatial memory and learning in Alzheimer-type dementia and Parkinson's disease. *Brain*, *111*, 695-718.

Saper, C. B., Iverson, S. & Frackowiak, R. (2000). Integration of sensory and motor function: the association areas of the cerebral cortex and the cognitive capabilities of the human brain. In: Kandel, E.R., Schwartz, J.H., Jessell, T.M. (Eds.), *Principles of Neural Science*. McGraw Hill, New York, 349-380.

Schultz, S. K., Ho, B. C. & Andreasen, N. C. (2000). Clinical features characterising young-onset and intermediate-onset schizophrenia. *Journal of Neuropsychiatry and Clinical Neurosciences*, *12*, 502-505.

Smith, C., Park, S. & Cornblatt, B. (2006). Spatial working memory deficits in adolescents at clinical high risk for schizophrenia. *Schizophrenia Research*, *81*, 211-215.

Tan, H., Choo, W., Fones, C. & Chee, M. (2005). fMRI study of maintenance and manipulation processes within working memory in first-episode schizophrenia. *American Journal of Psychiatry*, *162*, 1849-1858.

Tabachnick, B. G. & Fidell, L. S. (2001). *Using Multivariate Statistics*, fourth edition. Allyn and Bacon, Boston.

Thompson, P. M., Vidal, C., Giedd, J. N., Gochman, P., Blumenthal, J., Nicolson, R., Toga, A. W. & Rapoport, J. L. (2001). Mapping adolescent brain change reveals dynamic wave of accelerated gray matter loss in very early-onset schizophrenia. *Proceedings of the National Academy of Sciences*, USA, *98*, 11650-11655.

Tsuang, M. T., Stone, W. S., Tarbox, S. I. & Faraone, S. V. (2002). An integration of schizophrenia with schizotypy: identification of schizotaxia and implications for research on treatment and prevention. *Schizophrenia Bulletin*, *54*, 169-175.

Vance, A., Hall, N., Bellgrove, M., Casey, M., Karsz, F. & Maruff, P. (2006). Visuospatial working memory deficits in adolescent onset schizophrenia. *Schizophrenia Research*, *87*, 223-227.

Vance, A., Hall, N., Casey, M., Karsz, F. & Bellgrove, M. (2007). Visuospatial memory deficits in adolescent onset schizophrenia. *Schizophrenia Research*, *93*, 345-349.

Vidal, C., Rapoport, J. L., Hayashi, K. M., Geaga, J. A., Sui, Y., McLemore, L. E., Alaghband, Y., Giedd, J. N., Gochman, P., Blumenthal, J., Gogtay, N., Nicolson, R., Toga, A. W. & Thompson, P. M. (2006). Dynamically spreading frontal and cingulated deficits mapped in adolescents with schizophrenia. *Archives of General Psychiatry*, *63*, 25-34.

Wechsler, D. (1991). *Wechsler Intelligence Scale for Children*, third edition. The Psychological Co-operation, Texas.

Wood, S. J., Pantelis, C., Proffitt, T., Phillips, L. J., Stuart, G. W., Buchanan, J. A., Mahoney, K., Brewer, W., Smith, D. J. & McGorry, P. D. (2003). Spatial working memory ability is a marker of risk-for-psychosis. *Psychological Medicine*, *33*, 1239-1247.

Zakzanis, K. K. (2001). Statistics to tell the truth, the whole truth and nothing but the truth: formulae, illustrative numerical examples, and heuristic interpretation of effect size analyses for neuropsychological researchers. *Archives of Clinical Neuropsychology*, *16*, 653-667.

INDEX

A

B

C

F

G

J

K

L

M

N

O

P

Q

R

S

T